AF528454

Physico-chemical Parameters of Soil and Water

THE AUTHORS

Ms. Pappu Srinidhi was born in Madanapalle a town in Chittoor district which falls under state of Andhra Pradesh, India. Pappu Srinidhi obtained her Bachelor's degree in Agricultural science (B.Sc Agriculture) from Rai Technology University, Doddaballpura, Karnataka. She completed her post-graduation in Soil Science – Soil and Water Conservation from Banaras Hindu University, Varanasi, Uttar Pradesh. In her Post-graduation she worked on "Study of Physico-chemical Parameters of Soil and Water in Madanapalle block of Chittoor District, Andhra Pradesh" under the guidance of Dr. Y.V. Singh Associate Professor, Department of Soil Science and Agricultural Chemistry, Institute of Agricultural Sciences, Banaras Hindu University-Varanasi, Uttar Pradesh. Under his guidance she published 2 research papers in international journals. Pappu Srinidhi is currently Business Executive for RKVY-RAFTAAR project ANGRAU Poshan Incubator, Regional Agricultural Research Station, Tirupati, Andhra Pradesh, India.

Dr. Y. V. Singh originally belong to the Village Sujanpur, 32 km away from district head quarter of Agra, which falls under the state Uttar Pradesh. Dr Y.V. Singh graduated from R.B.S. College Bichpuri, Agra of Agra University Agra, (presently, Dr. Bhim Rao Ambedkar, University, Agra), also obtained his post graduate and Ph.D degree in Soil Science and Agricultural Chemistry discipline from the same campus. Dr. Y.V Singh started his professional carrier in the year 1996 and served various capacities in national and international projects up to 2006 at CSSRI Karnal, PDCSR, Meerut and IISS, Bhopal. Subsequent he joined the Department of SSAC, IAS, BHU in Nov, 2006 as a Assistant Professor and elevated to Associate Professor (Stege-4). He has been engaged in teaching, research and extension activities for more than 25 years. His field of specialization is Soil Fertility/STCR Correlation. He have received 19 awards of best paper presentation award, Excellence of teaching, best young scientist and excellence research appreciation from various organizations MRFSW Varanasi and VIF Chennai, SAID, Ranchi and PC, STCR, IISS, Bhopal. Presently, he is handling 2 research projects sponsored by CST, U.P. and STCR (AICRP) New Delhi as PI. Dr. Singh has successfully executed externally funded research projects: 3 as PI and 9 as Co-PI with active involvement in teaching of UG and PG students. He has guided 32 M.Sc. & 3 Ph.D. students. He has 220 publications including 90 peer reviewed research paper in reputed Indian and International Journals, 177 publications including 80 pear reviewed research paper (4 in Journal of Agrochimica, Field crop research and Soil Science Society of American) in reputed national and international journals, 40 popular articles, 16 full paper in conference proceeding, 09 technical bulletin. Dr Singh is currently Associate Professor (Stage-4), Department of Soil Science and Agricultural Chemistry, Institute of Agricultural Sciences, Banaras Hindu University – Varanasi (UP), India.

Physico-chemical Parameters of Soil and Water

Pappu Srinidhi
Y.V. Singh

2022
Scholars World
A Division of
Astral International Pvt. Ltd.
New Delhi – 110 002

© 2022 AUTHORS

ISBN : 9789354615184

Disclaimer:

Every possible effort has been made to ensure that the information contained in this book is accurate at the time of going to press, and the publisher and author cannot accept responsibility for any errors or omissions, however caused. No responsibility for loss or damage occasioned to any person acting, or refraining from action, as a result of the material in this publication can be accepted by the editor, the publisher or the author. The Publisher is not associated with any product or vendor mentioned in the book. The contents of this work are intended to further general scientific research, understanding and discussion only. Readers should consult with a specialist where appropriate.

Every effort has been made to trace the owners of copyright material used in this book, if any. The author and the publisher will be grateful for any omission brought to their notice for acknowledgment in the future editions of the book.

All Rights reserved under International Copyright Conventions. No part of this publication may be reproduced, stored in a retrieval system, or transmitted in any form or by any means, electronic, mechanical, photocopying, recording or otherwise without the prior written consent of the publisher and the copyright owner.

Published by : **Scholars World**
A Division of
Astral International Pvt. Ltd.
– ISO 9001:2015 Certified Company –
4736/23, Ansari Road, Darya Ganj
New Delhi-110 002
Ph. 011-43549197, 23278134
E-mail: info@astralint.com
Website: www.astralint.com

Acknowledgement

I bow my head before the feet of the "Lord Shiva", the 'Almighty' due to whose blessings I am able to reach at this stage and I feel that everything on this earth could be possible only with the blessings of this invisible but omnipresent supreme force.

At the outset, being the student of this great Institute, I bow my head with great reverence to the lotus feet of ***Mahamana Pandit Madan Mohan Malaviya Ji,*** the founder of the Banaras Hindu University, whose everlasting desire was to serve mankind. I am fortunate to perceive the prodigious path to tread upon precisely through precious guidance in this university.

I would like to express my profound sense of reverence and indebtness to my Supervisor, ***Dr. Y.V. Singh,*** Associate Professor (S-4), Department of Soil Science and Agricultural Chemistry, Institute of Agricultural Sciences, Banaras Hindu University for his meticulous guidance, compassionate initiation, congenial discussion, constructive criticism and soothing affection during the course of this investigation and preparation of this manuscript. It was a matter of sheer luck and opportunity to work under his guidance.

A special word of acknowledgement for ***Prof. Priyankar Rahat*** *and* ***Prof. Nirmal De,*** H.O.D (SWC) Soil Science & Agricultural Chemistry, Banaras Hindu University. Significant inputs from him have helped to enrich my work.

I offer my heartfelt gratitude to ***Dr. Ashish M Latare*** (Assistant Professor, Department of Soil Science and Agricultural Chemistry) R.G.S.C. (BHU) Barkkachha, Mirzapur, Institute of Agricultural Sciences, Banaras Hindu University for their constant encouragement, critical suggestions and inspiration during entire period of investigation and the members of the advisory committee ***Prof. P.K. Sharma*** Assistant Professor Department of Soil Science and Agricultural Chemistry Institute of Agricultural Sciences, Banaras Hindu University.

Dr. Pawan Kumar Anand (Horticulture), R.G.S.C. (BHU) Barkkachha, Mirzapur, ***Dr. R. Meena*** (Assistant Professor, Department of Soil Science and Agricultural Chemistry), Institute of Agricultural Sciences, Banaras Hindu University.

I am grateful to ***Prof. P. Rahat, Prof. A.K. Ghosh, Prof. Nirmal. De, and Dr. A. Rakshit.*** I am thankful to non-teaching staff members of Department of Soil Science and Agricultural Chemistry, Institute of Agricultural Sciences, Banaras Hindu University for all kinds of help to me during my study period.

My special thanks to ***Dr. Triyugi Nath, Dr. Savita Dewangan,*** for their guidance, construction criticism, affectionate behaviour and moral support during investigation of this research work.

I would like to thank sincerely from the bottom of my heart to my dearest friends ***Chichili jyoshna, Yogesh Chand Yadav, Debanjan Pal, Zion Adam Michael, Shubham, Gulshan, Anish, Kamal Kishore Patel, Ramveer*** for their kind encouragement and support during the conduct of research as well as while writing thesis.

Words with me are insufficient to express my feelings of my heart to acknowledge my gratitude to my beloved father ***Shri P.Sreekanth Reddy*** my mother ***Smt. P.Bhargavi Reddy.*** For their co-operation, inspiration and support in all the way and other family members who have providing all kinds of help and love.

Without the help of seniors none can learn the lesson of life and cannot teach the same to loving juniors. So, heartfelt and special thanks to my seniors for their co-operation during the study and investigation.

I would like to thank my juniors and others for always being supportive and helping and also thank to everyone who was involved in completing this work.

I would like to extend my sincere thanks to ***Sita Ram Yadav,*** Lab attendant, SS-SWC, RGSC, BHU. Last but not the least, I record my sincere thanks to all beloved and respected people who helped and could not find separate mentions. I still solicit their benediction to proceed at every step of respected destined life.

It's like drop in the ocean by my all regards to Maa Vindhyavasini and Baba Vishwanath, for providing me energy and patience without which I would have been none.

Pappu Srinidhi

Contents

Abbreviations

%	Per cent
Agri -	Agriculture
BD	Bulk density
Bis	Bureau of Indian Standard
Ca	Calcium
Cm	Centimeter
Cm^{01} (P^{+}) Kg^{-1}	Centimol per kilogram
DSm^{-1}	Desisimen per meter
DTPA	Di-ethylene Triamine Penta acetic acid
ET. al-	Et alia and others
G Kg^{-1}	Gram per kilogram
Ha.	Hectare
i.e.	Id.est. that is
J	Journal
K	Potassium
Kg ha^{-1}	Kilogram Pre hectare
Mea/100g	Milli eaquivalent Per 100 gram of soil
Mg	Magnesium
Mg Kg^{-1}	Milligram per Kilogram
Mha	Million hectare
Mm -	Millimeter
N	Nitrogen
Na	Sodium

NIV -	Nutrient index value
OC	Organic carbon
P	Phosphorus
P.D	Particle density
pH -	Puissance de hydrogen
S -	Sulphuric
Sq. Km	Square Kilometer
Viz.	Vide licet namely
WHC	Water holding capacity
WQI -	Water Quality index

List of Tables

List of Figures

1

CHAPTER

Introduction

Soil is a natural resource found on the surface of the earth. Soils are living, breathing world supporting nearly all terrestrial life on earth, but still most people do not realize it and yet they are taken for granted. Soils and its functions within an ecosystem vary from one location to the other location due to different factors which includes climate, plant and animal life living on them i.e. the decayed matter, parent material of the soil, topography i.e. soil position on landscape and age of the soil. The thin layer on the earth surface where soil develops is known as Pedosphere. This thin layer act as a "great integrator" that serves as a great natural resource and affects every part of ecosystem. Soil holds the nutrients and water for plants. They act as a filter to clean the water passing through them. Whatever the food we consume and materials we use like paper, clothing etc. are dependent on soil. They act as major resource for storing and transferring of the heat to the atmosphere thereby regulating the temperature. By studying all these functions and roles that soil play the students and scientists they get a keen interest to interpret a site geology, hydrology climate and vegetation so therein lies our interest to analyse and learn about different soils in an area (**Chesworth, 2008**).

Soil health is the other term which is used to denote the productive farming practise. Soil health and soil quality both the terms are interchangeable. Soil health, also referred to as soil quality, is defined as the continued capacity of soil to function as a vital living ecosystem that sustains plants, animals, and humans. This definition speaks to the importance of managing soils so they are sustainable for future generations (**USDA Natural Resource Conservation Service**). Whereas the other term which includes the soil fertility is the ability of the soil to provide nutrients to the plant for their growth and development. Generally soils are supplemented with the fertilizers, manures and composts in order to meet their nutritional requirements. There are 16 essential nutrient elements that are required for the plant growth absence of any one nutrient will render the crop barren. So there in necessary to analyse the soil to know the deficient nutrients and to add it in right quantity. The nutrients taken up by the plant are classified into primary nutrients, secondary nutrients and micronutrients. Water, oxygen and carbon dioxide are taken up by the plants from the atmosphere.

The primary nutrients include nitrogen, potassium and phosphorus these are the major nutrient fertilizers which are applied to the plants in large amount compared to secondary and micronutrients. A nutrient is said to be essential if it is directly involved in metabolism and plant cannot complete its life cycle in absence of that particular nutrient.

Calcium, magnesium and sulphur are known as secondary macronutrients which are taken up by the plants in smaller amounts for their growth and metabolism. Dolomite, limestone and calcite are the major source of calcium and magnesium deposition in soil. These minerals contain calcium and magnesium in the form of calcium oxide and magnesium oxide. Along with these a different form of fertilizers are also available. Potassium sulphate, magnesium sulphate, gypsum and elemental sulphur are different sulphur containing fertilizers.

Boron, chlorine, copper, iron, manganese, zinc and molybdenum are known as micronutrients which are required for the plants in very trace amounts. Different form of fertilizers are also available such as copper sulphate, manganese sulphate, zinc sulphate, zinc oxide and oxysulphates. Chelates are soluble organic compounds that form complexes with metals by increasing their solubility and making them available to plant roots these micronutrients are applied to soil in chelated forms. **(Allen *et al.* 2007).**

Apart from the soil quality the quality of irrigation water applied to the plant also determines its effect on growth and development. Generally quality of irrigation water describes the concentration and state of the organic and inorganic materials suspended in water. The determination of the quality is done in In-situ measurements and also by examining the samples in laboratory. The water used for irrigation can vary greatly depending on the type and the quantity of the salts dissolved. Generally the salts in the irrigation water originate from different sources which include weathering of the rocks, dissolution lime, gypsum and other minerals, sea water intrusion, etc. when such water is applied to the plant there will be an increase in osmotic potential and make it difficult for the roots to absorb the water from the soil and also as the water evaporates the salts get accumulates on surface of the soil as the white crust. Therefore water quality testing plays an important role in monitoring the plant life and also forms the important part of the environmental monitoring. The properties of the water which are determined for its quality are classified into chemical physical and biological properties. For irrigation waters the general criteria includes salinity, sodicity and element toxicities.

In Andhra Pradesh there are 6 major soil groups found these includes red soil belonging to order alfisols contribute about 65%, black soils belonging to order vertisols contribute about 25%, alluvial soils belonging to order entisols contribute about 5%, coastal sands belonging to order entisols contribute about 3% and laterite and lateritic soils belonging to order oxisols contribute about 1%.

Chittor is a district in Rayalaseema region covering an area about 15,359 square kilometres and divided into three revenue divisions namely Chittor, Tirupathi and Madanapalle. To its north the district is bounded by Kadapa district, to its south it is bounded by Krishnagiri district, Vellore District, Tirupattur district and Tiruvallur district of Tamil Nadu state, to its northwest it is bounded by Nellore district and Kolar district of Karnataka state to its northeast. Chittor district is present at extreme south and lies between 12°37′ - 14°8′ north latitudes and 78°3′ - 79°55′ east longitudes. The district is covered by forests about 30%. It famous for mango growing mainly in Satyavedu, Puttur and Srikalahasdti mandal. The soils are red loamy 57%, red sandy 34% and the remaining 9% is covered by black clay black loamy, black sandy and red clay. It is a major market Centre for mangoes, grains, sugarcane, and peanuts.

Commonly we can see granite and gneiss as major landscape about 36% and basalt accounting for 32% followed by laterite, alluvium, shale and sandstone. The area consists of very gentle to gently sloping lands (40%) followed by gently to moderate sloping lands (32%) and level lands of (10%). If we take a look out physiography upper Pedi plains occupies 40%, pediment 32% and lower Pedi plains 1.7%. Agriculture occupies about 32% where plantation accounts for 16% and scrubs 15%. Around 70% of the land constitutes for dry land agriculture and 30% of the land constitutes for Rainfed agriculture. 23 to 43% of the land is under moderate to severe erosion where 14% is under slight runoff potential. According to land capability classification 49.7 percentage of the area is under Class-III and 9.9 percentage under class II. 30 % of the area is classified under severe soil limitation for irrigation and 27 % of area is under marginal lands for sustained use under irrigation because of very severe limitations. As it is a part of Eastern Ghats the district is surrounded by hills and we can see a large number of chains of small tanks which were built during the Vijayanagar kings and this tanks form a backbone of irrigated agriculture in absence of canal or river projects.

By considering all these features an effort has been done to evaluate and describe the soil and water quality status of Madanapalle town in Chittor district, Andhra Pradesh by using different principles, procedures and techniques with the following objectives:

1. To evaluate the Physico-Chemical parameters of the soil and water collected from Madanapalle town of Chittor district Andhra Pradesh.
2. To study the fertility status of the soils present in Madanapalle block and prepare nutrient wise map in order to know the nutrient deficient areas.
3. To guide the farmers about the problems related to soil nutrients and water quality and techniques to overcome it.
4. To study the correlation between the soil and water parameters of Madanapalle block.

2

CHAPTER

Review of Literature

Soil and water are two major components important for plant growth metabolism and crop production. These components are related to each other i.e. change in one system causes varying changes in other. So the management practices to improve the crop production has to be done according to the changes done in a system. Based on the sources available and information gathered literature is reviewed under following sections.

2.1 Physico-chemical properties of soils

- Bulk density
- Particle density
- Porosity
- Water holding capacity
- Soil pH
- Electrical conductivity
- Organic carbon

2.2 Primary Macro-nutrients in soil

- Nitrogen
- Phosphorus
- Potassium

2.3 Secondary Macro- nutrients in soil

- Sulphur
- Calcium
- Magnesium

2.4 Physical and chemical properties of water

- Water pH
- Electrical conductivity
- Chloride
- Carbonate and Bicarbonate
- Calcium and magnesium
- Sodium and potassium

2.5 Irrigation water quality parameters

- Sodium adsorption ratio
- Residual sodium carbonate
- Soluble sodium percentage
- Permeability index
- Kelly's ratio

2.1. Physico-chemical properties of soils (pH, EC, Bulk density, Particle Density, porosity, WHC and organic carbon)

Chaudhari *et al.* (2013) studied the bulk density in relation to other soil physico chemical parameters under calcareous soil, black soil and red non-calcareous soil of Coimbatore district. In this study the samples were collected from eight different locations at a depth of 15cm. The results showed that the bulk density of the soil ranged from 1.25-1.57 Mg/m^3 and it is also observed that bulk density is dependent on calcareous soil and saline soil but independent on soil pH. There was a significant negative correlation of bulk density with calcium carbonate ($r = -0.4952$) and electrical conductivity ($r = -0.6611$).Whereas correlation between bulk density and pH was negative but not so significant ($r = -0.2371$).

Twum and Annang (2015) studied the impact of soil compaction on bulk density and root biomass of Quercus petraea *L.* at reclaimed post lignite mining site this study was conducted in Germany. The research was carried out by selecting two sites viz: compacted site and uncompacted site and at each site two sessile oak trees were selected for reference with comparable height and morphology. Bulk density was determined by using core method where samples were taken from compacted and uncompacted sites at 20-25cm, 100-105cm and 200-205cm.The results showed that bulk density was significantly affected by soil compaction. At the depths of 20 to 25cm and 100 to 105cm the compacted soil showed average value of 1.74 ± 0.09 g/cm^3 and 1.65 ± 0.06 g/cm^3 respectively which are high compared to uncompacted soils and at the depth of 200 to 205cm there was no significant difference between bulk density of compacted soil and uncompacted soil.

Ahad *et al.* (2015) studied soil bulk density as related to texture, organic matter content and porosity in Kandi soils of district Kupwara (Kashmir valley), India. In this study the soil samples were collected from five different locations at 20-35cm depth. The bulk density of the soils ranged from 1.24 – 1.46 gm/cm$^{3.}$ There was a strong negative correlation of bulk density with organic matter ($r = -0.87$) and porosity ($r = -0.79$).

Pandey *et al.* (2018) studied assessment of soil physical properties under different land use systems in a Mollisol at G.B Pant University, Pantanagar. The research was carried out under different land use systems viz: rice–wheat–green gram, rice–pea (vegetable)-maize, rice-potato –okra, rice–berseem + oat + mustard (fodder)-maize + cowpea (fodder), maize–wheat–cowpea, sorghum (fodder)-yellow sarsonblack gram, guava + lemon, poplar + turmeric, eucalyptus + turmeric and fallow (uncultivated land) and soil samples were collected at 0-20cm depth. The results showed that the particle density of a soil ranged from 2.55-2.74 g/cm^{3,}

bulk density ranged from 1.29-1.43 g/cm^3, porosity ranged from 47.62-49.71% and water holding capacity ranged from 41.56-61.82% which indicated that soils under agroforestry systems were superior followed by field crops, horticultural crops and uncultivated land.

Deb *et al.* (2013) studied physical and chemical properties of soils and its relationship with water holding capacity along altitudinal gradient for cultivated soils of South Sikkim. The study was conducted on a hilly terrain closed spaced contour with steep slope. A total of 35 soil samples were collected from a plough layer of 0-20cm depth from different locations of cultivated fields. The results showed that the bulk density of soil samples ranged from 1.08-1.54 gm/cc, particle density ranged from 2.05-2.77 gm/cc, porosity ranged from 42.22-61.73% and water holding capacity ranged from 33-73.86%. There was a significant positive correlation between water holding capacity and porosity (r = 0.5717) whereas negative correlation between water holding capacity and bulk density (r = -0.2677).

Viji and Rajesh (2012) studied assessment of water holding capacity of major soil series of Lalgudi, Trichy, India. Soil samples were collected from five different locations at 15cm and 30cm depths. The results showed that at 15cm depth the bulk density of soil samples ranged from 1.19-1.42 g/cm^3, particle density ranged from 1.46-1.79 g/cm^3, porosity ranged from 17.98-22.16% and water holding capacity ranged from 30.69-20.88%. Whereas at a depth of 30cm the bulk density of soil samples ranged from 1.21-1.41 g/cm^3, particle density ranged from 1.45-1.74 g/cm^3, porosity ranged from 15.92-19.60% and water holding capacity ranged from 20.46-29.33%.

Pandey *et al.* (2018) studied Physico-chemical characterization of Oak, Pine and Sal forest soil profiles of Betalghat Region of Kumaun Himalaya. In this study the soil samples were collected from three dominant forest types i.e. banj-oak, chir-pine and sal forest from three different depths viz: 0-10cm, 11-20cm and 21-30cm. The results showed that bulk density of soil samples ranged from 1.25-1.43 gm/cc and water holding capacity ranged from 33.90-69.27%.

Shivanna and Gopalpur (2014) studied chemical analysis of soil samples to evaluate soil fertility in Tiptur Taluk Karnataka, India. In this study five representative model plots were selected i.e. two plots from Halkurke tank command area, two from Eachanur tank command area and one from V. Mallenahalli tank command area and from each plot 10 point were selected in zigzag manner and soil samples were taken from two different depths of 0-15cm and 15-30cm. The results showed that pH of the soil samples from two sites of Halkurke command area were 7.07 and 7.54. pH values of the soil samples from two sites of Eachanur tank command area were 7.87 and 7.84. pH value of V. Mallenahalli tank command area was 7.52 and results showed that samples were alkaline in nature.

Jain *et al.* (2014) studied Physico-chemical characterization of farmland soil in Lunawada Taluk, Gujarat, India. In this study 15 soil samples from different villages were selected and analysed for Physico-chemical parameters. The results showed that pH of the soil samples ranged from 6.5-7.8, electrical conductivity ranged from 0.3-1.3 dsm^{-1} and organic carbon ranged from 0.3-0.8%.

Vijaya Kumar *et al.* (2011) studied and analysed 100 soil samples of tsunami affected areas of Sirkali Taluk in Tamil Nadu. The soil samples were collected from 0-15cm depth from 100 sites covering 18 villages. The results showed that pH of the soil samples ranged from 6.78- 7.95, electrical conductivity ranged from 0.07-0.62 dsm^{-1} and organic carbon content ranged from 0.20-7.54%. Organic carbon was found to be low in 64% of the samples due to the reason that vegetation was in poor condition and organic matter decomposition was at higher rate due to hyper-thermic temperate regime leading to high oxidation condition.

Das and Poongothai (2014) studied Soil and Water Quality in Tsunami affected Areas of Nagapattinam District, Tamilnadu, India. In this study 10 surface soil samples of unaffected area from 10 different villages were collected upto a depth of 20cm. The results concluded that the pH of a soil samples ranged between 6.82-8.60 and electrical conductivity ranged from 0.327-1.250. There was a decrease in electrical conductivity of the soils in comparison with the values immediately after tsunami-2004 this may be due to rainfall flushing out the surface salts thereby lowering soil EC.

Basavaraja *et al.* (2017) carried out research in Hassan district of Karnataka where he collected soil samples from eight different Taluks. Out of 2559 villages 220 villages were selected from Hasan district and from each village 6 soil samples at a depth of 0-15cm were collected representing 2 samples from small, medium and large farmers. The results stated that the pH of soil samples ranged from 3.72-9.71 and concluded that high pH has been observed due to high rainfall which caused leaching of bases. Due to leaching of bases the electrical conductivity of the soils were normal which ranged from 0.02-1.79 dsm^{-1}.

Kadam *et al.* (2016) studied Physico-chemical properties of soil in Deulgaon Raja region, Maharashtra. In this 60 soil samples were collected from different villages nearby Deulgaon Raja region. It was observed that soil type was clay loamy type the pH of the soil samples ranged between 5.20-6.8 where it was found 1.37% of the soil samples were acidic in nature and 92.64% of soil samples are in normal range. The EC of soil samples ranged between 0.30-0.46 dsm^{-1} and organic carbon content range was very high which ranged from 2.15-3.66%.

Chandak *et al.* (2017) studied Physico-chemical parameters of soil from Kadi city of Gujarat state. In this study 20 surface soil samples from five different regions were collected and stated that pH of soil samples ranged from 7.4-7.9 indicating alkaline in nature while EC of soils ranged from 0.3-0.6 mS/cm due to high rainfall which resulted in leaching of bases on soil surface.

Kumar *et al.* (2017) carried out research in Mid-Himalayan region of Himachal Pradesh on soil fertility status. The study area covered about 192.31 ha where 32% of the soil samples were collected from irrigated areas while the rest were collected from the fields depending upon the seasonal rains. The results stated that pH of the soil samples ranged from 6.59 – 7.81 due to high rainfall, lower temperature and higher altitudes. EC of soils ranged from 0.049-0.793 dsm^{-1} and organic carbon content ranged from 0.15-1.98% out of total soil samples only 11.6% of samples showed lower organic carbon whereas 31.6 and 56.8% of soil samples were in medium and high range.

2.2. Primary Macro-nutrients in soil (Nitrogen, Phosphorus and Potassium)

Marimuthu (2015) studied nutrient analysis of soil collected from Panuchakuli village, Kanyakumari district, Kanyakumari. In this study the soil samples were collected at 15cm depth for an acre and analysed for Physico-chemical parameters. The results concluded that the nitrogen content of the soil sample was 133.5 kg/acre, phosphorus content was 14.98 kg/acre and potassium content was 117.6 kg/acre.

Chandak *et al.* (2017) analysed Physico-chemical parameters of soil from Kadi city, Gujarat state. In this study the soil samples were collected from four different villages and from each village five soil samples were collected. The value of phosphorus ranged from 7.77 kg/ha to 23.31 kg/ha whereas potassium value ranged from 188.48 kg/ha to 243.04 kg/ha.

Jain *et al.* (2014) studied about the soils in Lunawada Taluka of Gujarat state. The soil samples were collected from 10 different villages and analysed for physico-chemical parameters. The results showed that value of nitrogen ranged from 0.03% - 0.07%, phosphorus ranged from 12 – 100 kg/ha whereas potassium value ranged from 132 – 914 kg/ha.

Wagh *et al.* (2013) studied Physico chemical analysis of soil from eastern part of Pune city. The study was conducted by collecting the soil samples from the sugarcane field in a period of 2009-2010 from Manjari, Hadapsar and Phursungi village situated at south east of Pune. Total of 12 soil samples were collected from each village. The results stated that the phosphorus content of soil samples ranged from 10-172.9 kg/ha this may be due to excessive use of phosphatic fertilizers in the fields whereas potassium content ranged from 112-840 kg/ha indicating sufficient K in most of the sample.

Chaudhari (2013) studied physico-chemical parameters of soil in Jalagon District, Maharashtra state. In this study soil samples were collected from eight different villages during the summer season. The results showed that nitrogen content of soil samples ranged from 0.036% - 0.049%, phosphorus content ranged from 0.023% - 0.036% whereas potassium content ranged from 0.89% - 1.39%. According to these results it is justified that all three macro nutrients are in normal range.

Kumar *et al.* (2017) carried out research in Mid-Himalayan region of Himachal Pradesh on soil fertility status. The study area covered about 192.31 ha where 32% of the soil samples were collected from irrigated areas while the rest were collected from the fields depending upon the seasonal rains. The results shown that nitrogen in soil samples were in a range of 201.2 – 603.3 kg/ha with a mean value of 312.9 kg/ha. In this study there was a positive correlation between nitrogen and organic carbon. The phosphorus in soil samples were in a range of 7-45 kg/ha. Almost 2.8 i.e. very less percentile of soils showed phosphorus deficiency. This is due to low availability of calcium, magnesium and sodium oxides at high altitudes that restrict phosphorus for fixation and precipitation. The potassium in soil samples were in a range of 70.5-448 kg/ha with a mean value of 193.8 kg/ha. Among all soil parameters there was a significant positive effect showed by organic carbon on availability of N, S, P, B and Mn whereas there was strong positive correlation between nitrogen and organic carbon (r = 0.351**), nitrogen with phosphorus and phosphorus with potassium.

Dameshwar *et al.* (2018) carried out research on soil fertility status of Kasdol block in Matya village, Chattisgarh. In this study the soil samples were collected from 15-20cm depth across a grid from 10ha arid area, Rainfed area whereas 2.4ha for irrigated area within each of such sample point five samples were collected randomly. The results conclude that available nitrogen in soil samples were in a range of 99.47- 281.4 kg ha^{-1} with an average of 152.8 kg ha^{-1}. Phosphorus in samples were in a range of 4.75 – 18.5 kg ha^{-1} with an average of 10.4 kg ha^{-1} and Potassium in soil samples were in a range of 145.0-585.5 kg ha^{-1} with an average value of 280.4 kg ha^{-1}.

Singh *et al.* (2018) studied Soil Fertility Evaluation for Macronutrients in some soils of Varanasi district of eastern Uttar Pradesh, India. In this study 35 surface soil samples were collected from 15 villages from Varanasi district at a depth of 0-15cm. The results stated that the nitrogen content of the soil samples ranged from 158.4-455. Kg/ha where 40% of the soil samples were at low nitrogen content. Phosphorus content of soil samples ranged from 133.4-17.10 kg/ha i.e. 90% of soil samples are in medium to high grade and 10% are in low grade. Whereas potassium content ranged from 157.42-122.40 kg/ha i.e. 63% of the soil samples are in medium grade and 37% of the soil samples are in low grade.

2.3. Secondary Macro- nutrients in soil (Calcium, Magnesium, Sulphur)

Chaudhari (2013) studied physico-chemical parameters of soil in Jalagon District, Maharashtra state. In this study soil samples were collected from eight different villages during the summer season. The results stated that the percentage of Calcium in soil samples ranged from 0.94 to 3.38 where normal range of calcium is 0.98-2.45% by weight according to the study in sample S6 the percentage of Ca was 3.38% by weight which is abnormal. The percentage of magnesium in soil samples ranged from 0.91 to 2.65% by weight where the normal percentage of magnesium is 1.20 to 2.00% by weight, according to the study in sample S8 the percentage of Mg was 2.65% by weight which is abnormal.

Soni (2016) studied Physico-chemical parameters of soil samples from Abohar city. In this study the five soil samples were collected from a depth of 0-20cm from different places of the city i.e. 1 to 2 samples from central Abohar, 3 to 4 samples from western Abohar and 5th sample from eastern Abohar. The results stated that the sulphur content in soil samples ranged from 0.145g to 0.563g.

Gothwal and Kumar (2018) studied Physico-Chemical Analysis of Soil during Summer Season in Lentic Fresh Water Ecosystem Nakki Lake-Mount Abu (Rajasthan), India. In this study the soil samples were collected from three respective sites which included mountainous zone, populated zone and unpopulated zone from a depth of 15-20cm. The results concluded that the magnesium content of the soil sample from a mountainous zone was 150 mg/l, magnesium content from populated zone was 158 mg/l whereas from an unpopulated zone the magnesium content was 153 mg/l. The sulphur content of the soil sample from a mountainous zone was 245 mg/l, sulphur content from populated zone was 223 mg/l whereas from an unpopulated zone the sulphur content was 265 mg/l.

Singh *et al.* (2017) carried out research on physico-chemical characteristics of the soil of Lahar block in Bhind district of Madhya Pradesh. In this study ten surface soil samples were collected from farmer's field at a depth of 0-15cm in Lahor block, Bhind district. The results stated that the calcium content in the soil samples ranged from 4.50-9.05, magnesium content in soil samples ranged from 2.73-6.99 whereas sulphur content of the soil samples ranged from 16.58-26.68. According to the results calcium and magnesium were found 100% sufficient in the soils and sulphur was found to be medium in 40% of samples and high in 60% of the samples.

2.4. Physical and chemical properties of water

➢ pH and EC

Gothwal and Kumar (2018) studied lentic fresh water ecosystem during summer season in Nakki-Lake, Mount-Abu, Rajasthan. In this study three water samples were collected from three different zones i.e. populated zone, non-populated zone and mountainous zone. The results stated that the pH and TDS of populated zone was 7.11 and 142 mg/L, pH and TDS of non-populated zone was 6.95 and 146 mg/L whereas pH and TDS of mountainous zone was 7.11 and 193 mg/L.

Chakraborty and Nandini (2015) studied drinking water quality assessment for pH and total dissolved salts in Bangalore. In this study the whole city was divided into six divisions i.e. North, East, South East, South, West and Central Bangalore. Total of 1581 water samples were collected and analysed for pH and TDS. The results showed pH of the water samples ranged from 5.0-9.0 where 77.86% of the samples are in permissible limit and TDS of the water samples ranged from 10-1500ppm where 82.66% of the samples are below the permissible limit.

Rahman *et al.* (2017) studied ground water irrigation quality in South-Central Bangladesh. In this study a total 46 ground water samples were collected in two seasons i.e. wet monsoon season and dry monsoon season. The results stated that during dry and wet monsoon periods the pH ranged from 4.5-11.0 and 7.20-7.9 whereas EC during dry and wet monsoon periods ranged from 274-9190 and 600-9380 μs/cm.

Gummadi *et al.* (2015) studied irrigation water quality with reference to coastal Andhra Pradesh, India. In this study 20 villages from Bapatla were selected for collecting the water samples and the water samples were analysed for pH, EC, manganese and iron. The results concluded that the pH of water samples ranged from 7.54-8.52 and the mean value of electrical conductivity was 577.7 μmohs/cm.

Kumar and Kumar (2010) studied water quality parameters of ground water in Ambattur town, Tamil Nadu. In this study water samples were collected from 10 different sites during post rainy season. The results of the study showed that the pH of water samples ranged from 7.2-8.5 and it was concluded that the samples were alkaline in nature whereas TDS of water samples ranged from 450-890 mg/l and it was concluded that TDS was in a permissible limit based on WHO standards.

Kumar *et al.* (2014) carried out research on shallow hard rock aquifer to study ground water quality of Pudunagaram, Palakkad district, Kerala. In this a total of 60 ground water samples were taken during post-monsoon and pre-monsoon period. The results stated that during post- monsoon the pH and EC of water samples ranged from 6.1-7.8 and 3000-107 mg/L whereas during pre-monsoon pH and EC of water samples ranged from 6.9-7.3 and 3380-178 mg/L. The EC values during pre-monsoon period shown that 93 % of wells are out of limit whereas in post monsoon season 86% of the wells were in out the limits this reported due to some anthropogenic activities carried during pre-monsoon season.

Mayas *et al.* (2015) carried out research at Maytam area in city of Ibb, Yemen to study ground water quality. In this study 10 water samples were selected from different boreholes where one borehole is located at the upstream of the waste water treatment plant, another bore hole is located very close to waste water treatment plant, while others are located at the downstream of the waste water treatment plant. The results showed that the pH value ranged from 7.00-7.8 with an average value of 7.37, indicating alkaline nature of the ground water samples. Electrical conductivity of the water samples ranged from 1079-1456 μs/cm.

Mehta *et al.* (2014) studied hydro geochemistry of ground water in Sanganer Tehsil, Jaipur district, Rajasthan. In this study a total of 40 samples were collected from wells, tube-wells and hand pumps during pre-monsoon period. The results concluded that pH of water samples ranged from 8.87-7.25 where the water is alkaline in nature and showed negative correlation with all other parameters. EC of water samples ranged from 4850-530 μmhos/cm and all samples were higher than the permissible limit.

Nag and Das (2014) studied groundwater for irrigation and domestic suitability in Suri 1 and 2 blocks, Birbhum district, West Bengal. In this study 26 ground water samples were collected from boreholes of depth 2-21m during pre-monsoon and post-monsoon period. The results showed pH and EC of water samples during post-monsoon ranged from 7.0-8.4 and 90-300 μs/cm whereas during pre-monsoon the pH and EC of water samples ranged from 8.4-6.4 and 55-590 μs/cm.

Reda (2016) studied Physico-Chemical analysis of drinking water quality of Arbaminch town. In this study the water samples were collected from three sampling sites i.e. Secha, Sikela and Arbaminch University. The results showed that pH of three sampling sites i.e. Secha, Sikela and Arbaminch University were 7.34, 7.13 and 7.76 and stated that all samples are in permissible range according to WHO standards whereas EC of three sampling sites i.e. Secha, Sikela and Arbaminch University were 211.8, 210.0 and 230.5 μs/cm and stated that all samples are in permissible range.

- **Anions (Cl^-, CO_3^{2-} and HCO_3^-)**

  Shukla and Arya (2018) studied chloride ion concentration in Ganga river water by Mohr method at Kanpur, India. In this study water samples were collected from three different sites (Bithor ghat, Siddnath ghat and Dhoni ghat) during pre-monsoon, monsoon and post-monsoon of 2016-2017 and 2017-2018. The results showed that in the year 2016-2017 pre-monsoon period the chloride concentration in three sites i.e. Bithor ghat, Siddnath ghat and Dhoni ghat was 43.96, 113.44 and 82.24 mg/L, during monsoon period the chloride concentration was 31.20, 48.21 and 34.67 mg/L and during post-monsoon the chloride concentration was 61.06, 92.3 and 63.9 mg/L. Whereas during the year 2017-2018 pre-monsoon period the chloride concentration in three sites i.e. Bithor ghat, Siddnath ghat and Dhoni ghat was 50.23, 120.12 and 84.05 mg/L, during monsoon period the chloride concentration was 34.67, 51.40 and 41.23 mg/L and during post-monsoon the chloride concentration was 50.45, 80.01 and 62.83 mg/L.

Amanial (2016) studied Physico-Chemical analysis of drinking water quality of Arbaminch town. In this study the water samples were collected from three sampling sites i.e. Secha, Sikela and Arbaminch University. The results showed the chloride concentration in Secha, Sikela and Arbaminch University was 135.22, 136.58 and 554.6 mg/L respectively which stated that the chloride concentrations of Secha and Sikela are within the permissible limits as stated by WHO(250 mg/L) and BIS(200 mg/L) whereas chloride concentration in Arbaminch town was higher than the permissible limit.

Chaudhary and Kumar (2018) studied groundwater quality for drinking and irrigation in arid areas of Rajasthan, India. In this study 300 ground water samples were collected from privately owned hand pumps in canal catchment areas of site1: Indira Gandhi, site2: Bhakra and site3: Ganga during pre-monsoon period. The results stated that the concentration of bicarbonate ion in three sites i.e. Indira Gandhi, Bhakra and Ganga ranged from 589.87-203.13, 599.63-381.25 and 536.8-278.16 mg/L.

Khichi (2018) studied Narmada river water quality assessment in Rajghat, Madhya Pradesh. In this study a total of 20 water samples were collected from August 2011 to July 2012 and analysed for quality assessment. The results of the study concluded that alkalinity of the water ranged from 293-132 mg/L.

Acharya *et al.* (2010) studied Physico-Chemical parameters of irrigation water, Prantij, Gujarat, India. In this study total of 20 water samples were collected from bore wells and analysed according to standard methods recommended by APHA (1995). The results showed that bicarbonate concentration in samples ranged from 5.50-14.50 Meq/lit where all samples are below the permissible limit 120 Meq/lit whereas chloride concentration in water samples ranged from 10.4-3.2 Meq/lit where 53% of the samples are having higher chloride concentration than permissible limit.

Mayas *et al.* (2015) carried out research at Maytam area in city of Ibb, Yemen to study ground water quality. In this study 10 water samples were selected from different boreholes where one borehole is located at the upstream of the waste water treatment plant, another bore hole is located very close to waste water treatment plant, while others are located at the downstream of the waste water treatment plant. The results showed that bicarbonate and chloride concentration in water samples ranged from 395.13-687 mg/l and 76.67-212.33 mg/l respectively which stated all samples are in maximum permissible limit as per standards of WHO.

Kant *et al.* (2015) carried research on ground water quality of Lahar block, Bhind district in Madhya Pradesh. In this study 40 water samples were selected throughout the study area. The results concluded that chloride concentration in water samples ranged from 1.90-15.30 Meq/lit, carbonate concentration ranged from 0.0-1.1 Meq/lit whereas bicarbonate concentration ranged from 2.10-8.30 Meq/lit.

Nag and Das (2014) studied groundwater for irrigation and domestic suitability in Suri 1 and 2 blocks, Birbhum district, West Bengal. In this study 26 ground water samples were collected from boreholes of depth 2-21m during pre-monsoon and post-monsoon period. The results showed bicarbonate concentration in water samples ranged from 48.80 to 1073.60 mg/L in post-monsoon and in pre-monsoon the bicarbonate concentration ranged from 48.8-292.8 mg/L. The chloride ion concentration in water samples were in a range of 15.0 to 124.96 mg L^{-1} in post-monsoon and 15.0 to 274.9 mg L^{-1} in pre-monsoon.

Periyaswamy *et al.* (2015) studied irrigation water quality assessment in Tamiraparani River, Southern India. In this study water samples were collected from 20 different locations during four different seasons i.e. pre-monsoon, monsoon, post-monsoon and summer. The results stated that alkalinity concentration in water samples ranged from 15.1-256.2 mg/L during summer and constant during the monsoon seasons not exceeding 120mg/L. The chloride concentration in samples ranged from 8.9 to 106.4 mg/L during pre-monsoon, 7.7 to 65.2 mg/L during monsoon, 8.9 to 260.5 mg/L during post-monsoon and 7.6 to 150.7 mg/L during summer season.

Alper and Hakan (2016) studied assessment of water quality in drainage canals of Carsamba plain, Turkey during July 2012 to January 2013. In this study 21 drainage canals are selected and total of 42 water samples were collected and results concluded that the bicarbonate ion concentration in July 2012 was in a range of 192.4 – 1873.8 mg L^{-1} which showed large variation with an average of 489.8 mg L^{-1}. In January bicarbonate ion concentration was in a range of 54.29 to 483.73 mg L^{-1} with an average of 253.88 mg L^{-1} and chloride and sulphate ion concentrations were in a range of 15.37 to 490.48 mg L^{-1} and 263.37 - 9.35 mg L^{-1} in July 2012 and January 2013 respectively.

Cations (Na^+, K^+, Ca^{2+} and Mg^{2+})

Mendonsa and Vishnu (2019) studied ecological analysis of five fresh water lakes in and around Mumbai, India. In this study five lakes were selected for sampling from September 2018 to March 2019 viz: Karnala Lake, Kala Talao Lake, Shivaji Talao Lake, Adharwadi Lake and Jagannath Kasinath Salvi Talao Lake. The results showed that the hardness i.e. concentration of calcium and magnesium ranged from 96.1 to 28.05 mg/L in the month of March where as in the month of September hardness ranged from 88.17-24.04 mg/L.

Sanda and Dibal (2014) studied irrigation water quality at Kadawa irrigation project during 2013 and 2014 irrigation projects there were three treatments used for sampling viz: canal water, drainage water and a blend of irrigation/drainage water. The results concluded that in the year 2013 and 2014 in drainage water, irrigation water and drainage/irrigation water the calcium concentration was 2.55 and 9.03, 120 and 3.05, 1.48 and 6.87, respectively. In the year 2013 and 2014 in drainage water, irrigation water and drainage/irrigation water the magnesium concentration was 1.18 and 1.98, 0.54 and 0.73, 0.83 and 6.25, respectively.

Sharma *et al.* (2014) studied surface water quality for irrigation purposes in mining areas of Solan district, Himachal Pradesh, India. In this study surface water samples were collected from 15 different locations in Kashlog mining area in the month of May and October 2011. The results concluded that the sodium concentration during pre-monsoon and post-monsoon ranged from 8.88-3.48 and 8.67-3.33 % respectively.

Kumar *et al.* (2014) carried out research on shallow hard rock aquifer to study ground water quality of Pudunagaram, Palakkad district, Kerala. In this a total of 60 ground water samples were taken during post-monsoon and pre-monsoon period. The results stated that during post- monsoon the calcium concentration in water samples ranged from 219-5 mg/L and 776-20 mg/L during pre-monsoon. The sodium concentration during post-monsoon and pre-monsoon ranged from 88.2-51.5 and 388-10.8 mg/L respectively.

Mayas *et al.* (2015) carried out research at Maytam area in city of Ibb, Yemen to study ground water quality. In this study 10 water samples were selected from different boreholes where one borehole is located at the upstream of the waste water treatment plant, another bore hole is located very close to waste water treatment plant, while others are located at the downstream of the waste water treatment plant. The results concluded that in July 2012 calcium concentration was in a range of 99.07-26.51, magnesium concentration was in a range of 113.38-14.03, sodium concentration was in a range of 700.3-18.10 and potassium concentration was in a range of 30.2-3.07 whereas in January 2013 calcium concentration was in a range of 87.81-4.91, magnesium concentration was in a range of 92.44-3.80, sodium concentration was in a range of 168.76-5.66 and potassium concentration was in a range of 9.27-1.24.

2.5. Irrigation Water Quality Parameters (SAR, RSC, SSP, PI and KR)

Ali *et al.* (2009) studied suitability of ground water for irrigation in Lahore district. In this study water samples were collected from 20 different villages in Lahore (Dist.) Water samples from three tube wells from each village were collected during the month of October 2005. Tube well selection was made at random and depth of bores ranged from 60 to 200ft. The results showed that the SAR concentration ranged from 0.12 to 33.5 with mean of 8.69 whereas RSC value of collected water sample ranged from 0 to 21.47me L^{-1} with mean of 6.51 Meq L^{-1}.

Bhat *et al.* (2016) studied Ground water Quality of irrigation in Gohana block, Sonipat Haryana, India. In this study 81 ground samples were collected to cover entire study area. The SAR values ranged from 4.03 to 24.16 ($mmol^{-1}$) and However, none of the samples were found unsuitable for irrigation. The PI values of the groundwater samples varied from 57.27 to 148.99 with a mean of 90.55. KR values of groundwater varied from 1.11 to 13.14.

Mirza *et al.* (2012) studied Groundwater quality for irrigation of deep aquifer in Satkhira Sadar Upazila of Satkhira district, Bangladesh. In this study twenty samples of groundwater were collected and analysed for. The results showed salinity (0.65-4.79 ppt), sodium absorption ratio (1.14-11.62), soluble sodium percentage (32.95-82.21), electrical conductivity (614-2082.11 μS/cm), magnesium adsorption ratio (21.96-26.97), Kelly's ratio (0.48-4.62), total hardness (150.76-313.33 mg/l), permeability index (68.02-94.16) and residual sodium bi-carbonate (79.68-230.72 mg/l). Chemical constituents and values were compared with national and international standards. Northern deep groundwater has the highest salinity and chemical concentrations whereas salinity and other chemical concentrations showed a decreasing trend towards the south.

Singh *et al.* (2017) carried out research on physico-chemical characteristics of the soil of Lahar block in Bhind district of Madhya Pradesh. In this study ten surface soil samples were collected from farmer's field at a depth of 0-15cm in Lahor block, Bhind district. The results stated that the SAR values ranged from 0.87-28.22 which showed that 82.50% of area is suitable for irrigation and 17.50% of area falls under high to very high SAR for irrigation. Kelly's ratio ranged from 0.14-8.05 whereas SSP value ranged from 88.95-12.56.

3
CHAPTER

Materials and Methods

The present chapter dealt with materials and methods followed during research work in evaluation of physical and chemical parameters of soil and water collected from Madanapalle block. The information on the collected site, data regarding the soil and water and procedures and techniques followed during research had been briefly described under different sections and sub-sections.

3.1. Reagents and instruments used in soil analysis and water analysis

Table: 3.1a. Reagents and solutions used in soil analysis

ANALYSIS	REAGENTS USED
Soil pH	• Buffer solutions of 4.0, 7.0 and 9.2 • Soil and distilled water suspension in the ratio of 1:2.5
Soil electrical conductivity	• Soil and distilled water suspension in the ratio of 1:2.5
Soil nitrogen	• 2.5% of sodium hydroxide • Mixed indicator(0.1 of bromocrosol green and 100ml ethanol) • 2.5% of boric acid • 0.32% of potassium chromate
Soil phosphorous	ACIDIC MEDIUM • Bray's reagent(0.03N ammonium fluoride + 0.025N HCL) • Ascorbic acid • Molybdate tartrate solution(ammonium molybdate + antimony potassium tartrate + H_2SO_4) BASIC MEDIUM • O.5N sodium bicarbonate solution • Dacro-g-60 or charcoal • Ammonium molybdate solution • Ascorbic acid • Antimony potassium tartrate solution • 2.5M sulphuric acid • P-nitro phenol indicator
Soil potassium	• Standard potassium chloride solutions • 1N Ammonium acetate solution

ANALYSIS	REAGENTS USED
Soil calcium and magnesium	• Ammonium acetate solution • Aqua regia(HCL + HNO_3) • Erichrome black-T indicator • EDTA(ethylene diamine tetra acetic acid) • Muroxide indicator • 4N NaOH buffer solution • Ammonium chloride – ammonium hydroxide buffer solution
Soil organic carbon	WALKLEY AND BLACK RAPID TITRATION METHOD • 1N potassium dichromate • 0.5N ferrous ammonium sulphate • Diphenylamine indicator • Concentrated sulphuric acid(sp.gr.1.84) • Orthophosphoric acid or sodium fluoride
Soil sulphur	• 0.15% of calcium chloride • Barium chloride crystals • 0.25g of gum acacia • Standard potassium sulphate solutions

Table: 3.1b.Reagents and solutions used in water analysis

ANALYSIS	REAGENTS USED
Water pH	• Buffer solution of pH 4.0, 7.0 and 9.2
Water electrical conductivity	• 0.01N KCL solution
Carbonates and bicarbonates	• 0.05N standard sulphuric acid • Methyl red indicator • Phenolphthalein indicator
Calcium and magnesium	• 0.01N EDTA solution • Muroxide indicator • Sodium diethyl dithio carbamate crystals • 4N NaOH buffer solution • Erichrome black-T indicator • Ammonium chloride-ammonium hydroxide buffer
Sodium	• Standard sodium chloride solution
Chloride	• 0.02N silver nitrate solution • 0.02N sodium chloride solution • Potassium chromate indicator
Potassium	• Standard potassium chloride solution

Table: 3.1c. Instruments and apparatus used in soil and water analysis

INSTRUMENTS	
pH meter	For measuring soil and water pH
Electrical conductivity meter	For measuring salinity in soil and total dissolved salts(TDS) in water

INSTRUMENTS	
Hot air oven	For drying the soil samples for 24hrs @105°C
Flame photometer	For measuring the soil and water potassium, sodium & calcium content
Spectrophotometer	For measuring phosphorus content in soil
Water bath	To evaporate the contents
Shaker	For uniform mixing
Weighing balance	To measure the chemicals in grams or kilograms
Semi-Auto Nitrogen analyser	For estimation the available nitrogen in soil

3.2 Description of Site

3.2.1. Location

Madanapalle is a town in Chittor district of the Indian state of Andhra Pradesh. It comes under municipality of Madanapalle mandal and Madanapalle revenue division. The area shows the ecological features of Deccan Plateau, Eastern Ghats and Western Ghats as it is located at the tri-junction of these three ecological features. The area is surrounded with the broken hill ranges with forestlands on ridges, farmlands and waste lands which are found on the lower slopes with numerous irrigation tanks. There are total of 21 villages in this mandal.

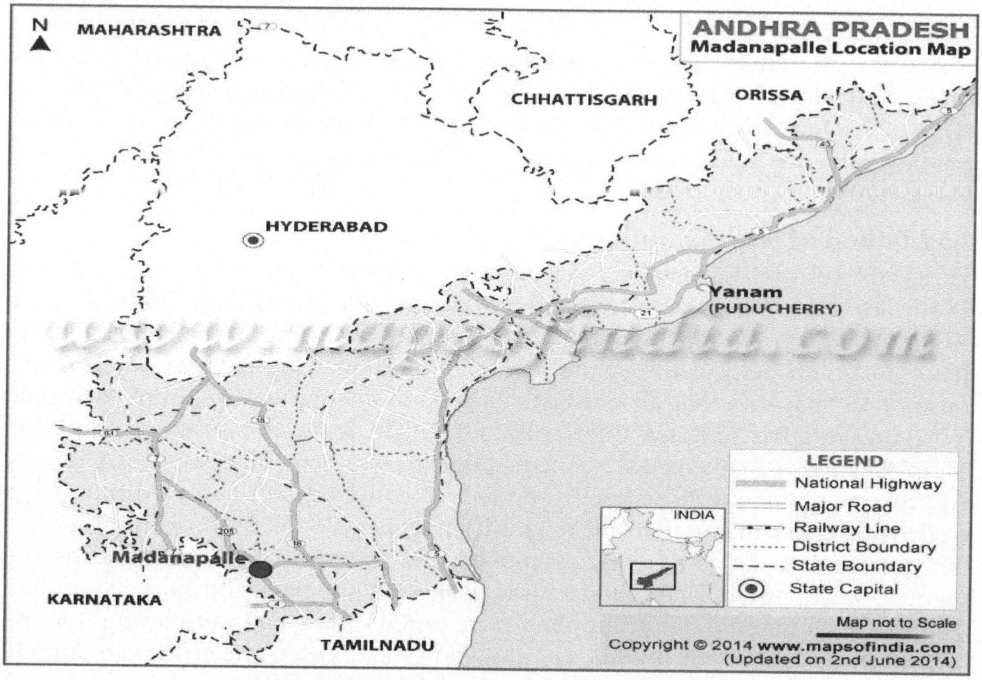

Fig. 3.2.1a: Location map of Madanapalle

3.2.2. Geography

Madanapalle is located at an average elevation of 695m i.e. 2,280ft above the mean sea level and it is located at 13.55°N and 78.50°E. The place shows tropical type of climate with pleasant mild warm summers with an average temperature of 30-35°C and during summers the temperature do not exceed to 40°C. Winters in Madanapalle are cold with temperature between 7-15°C. Summer season lasts from March to June with rainy season in June followed by winter which lasts upto February.

Table: 3.2.2a. Climate data of Madanapalle

Month	Jan	Feb	Mar	Apr	May	Jun	July	Aug	Sep	Oct	Nov	Dec
Average high temperature in °C	27.3	30.2	33.4	34.9	35	32.1	30.2	30.1	29.9	28.6	26.8	25.7
Average low temperature in °C	15.5	16.8	19.4	22.2	23.6	22.8	21.8	21.8	21.2	20.2	17.8	15.6
Average precipitation (mm)	4	2	3	28	61	51	81	73	111	143	54	32

(Source: Wikipedia.org)

3.2.3. Soils and Agriculture

The soils of Madanapalle varies from medium to deep red loamy soils and medium black soils. But in the district, red loamy soils constitute 57percent and red sandy soils 33percent. The remaining 9 percent is accounted for by black clay, black loamy, black sandy and red clay soils. The major economy of the town is based on agriculture. The place is known for its tomato growing and is known as Tomato hub of Andhra Pradesh. Madanapalle is the biggest tomato market in Asia. Tomato, groundnut, mango and tamarind are the major crops grown in Madanapalle town. Tomato in red gram, red gram in groundnut, horse gram in mango are the new intercropping systems found in Madanapalle town.

3.3 Method of Soil Sampling

Three soil samples were collected from eight different sites/villages i.e. a total of 24 surface soil samples were collected in Madanapalle Mandalam, Chittor district, Andhra Pradesh. At first twigs, litter and stones were removed from sampling place and with help of the spade the soil was digged upto 15cm depth in the form of V- shape so that all horizons of the soil are shown up. From each field 5-7 representative samples were collected in the form of zigzag manner. Soil samples which are collected are of 2cm in thickness, 3-5cm width and 15cm length. Approximately about 3kg of soil sample was collected and spread on the large cloth and mixed in order to make soil homogenous once the mixing was done left over root twigs, leaves were removed and the sample was reduced to 1kg by quadratic method. Then the soil was packed in jute bags and sample number, sample location and date of sampling were noted on the bags and brought to the laboratory. In laboratory the soil samples were shade dried and crushed manually with the help of mortar and pestle and sieved through 2mm size sieve mesh. About 500g of soil sample was obtained after sieving and thus obtained soil sample was stored in a clean jute bag. In this study 8 different villages were selected for soil and water sampling where land types were cultivated and fallow lands and total of 24 soil samples were collected and analysed in Rajiv Gandhi South Campus, Banaras Hindu University, Mirzapur, Uttar Pradesh. **Description of soil sampling sites is presented in Appendix-1.**

3.4 Physical and chemical analysis of soil samples

The soil samples collected were analysed for physical and chemical parameters in laboratory. The physical parameters include bulk density, particle density, porosity and water holding capacity whereas chemical parameters include pH, electrical conductivity, organic carbon, primary macronutrients and secondary macronutrients.

3.4.1 Determination of Water Holding Capacity

The determination of soil water holding capacity was done with the help of keen box which was suggested by **Piper (1966)**. A round filter paper was placed at the inner base of the keen box. The weight of keen box+filter paper was noted. The box was then filled with soil by tapping 20-30 times approximately by making plane at the top of the soil. Then a keen box containing filter paper and soil was placed inside the petridish containing water for 1hour so that the soil in keen box absorbs the maximum amount of water. After 1hour the keen box was placed on a filter paper to remove excess water. The keen box containing wet soil weight was recorded. Then the keen box containing wet soil was oven dried @ 105°C for 24hrs that oven dried soil weight was recorded.

$$\text{Saturation \%} = \frac{wet\ wt. - dry\ wt.}{dry\ wt.} \times 100$$

3.4.2 Determination of Bulk Density

The determination of soil bulk density was done with the help of Pycnometer which was suggested by **Black (1965)**. Firstly on weighing balance pycnometer weight which is of clean and dry was taken. Then soil was filled in the pycnometer upto to the brim by tapping and pycnometer and soil weight together was recorded. Then soil from pycnometer was removed and it was filled with water drop by drop with help of burette and how much ml of water has been used to fill the pycnometer by burette was taken as volume of soil. The reading will give the actual volume of pycnometer. Then the bulk density was calculated by using formula.

$$\text{Bulk density} = \frac{mass\ of\ the\ soil}{volume\ of\ the\ soil}$$

3.4.3 Determination of Particle Density

The determination of soil particle density was done with the help of Pycnometer which was suggested by **Black (1965)**. Firstly on weighing balance pycnometer weight which is of clean and dry was taken followed by pycnometer + water weight was taken. Then in a beaker 10g of soil and 15ml of water was added and boiled for few minutes in order to expel the air. Then the boiled contents were transferred into the pycnometer and filled with water upto the brim and weighed along with stopper. Using these values particle density was calculated

$$\text{Particle density} = \frac{10}{(W2+10)-W3}$$

- Mass of empty, clean, dry pycnometer (g) = W_1
- Mass of pycnometer + water (g) = W_2
- Mass of pycnometer + water + soil (g) = W_3
- Mass of soil (Ws) = 10g

3.4.4. Determination of porosity

Porosity was calculated by using bulk density and particle density values of the soil by using the formula

$$\text{Porosity (\%)} = \left[1-\frac{\textit{bulk density}}{\textit{particle density}}\right]\times 100$$

3.4.5. Determination of organic carbon

Organic carbon was estimated by wet oxidation method as suggested by **walkley and black (1934)**. In a conical flask 1g of soil sample was taken followed by 10ml of potassium dichromate solution was added. Then 20ml of sulphuric acid was added to the sample there was a change in the colour to green which indicates the presence of the organic matter and left it for 30min for oxidation. Then 200ml of distilled water was added to the samples and 0.5-1ml of diphenylamine indicator was added to the samples. There was a change in colour to purple this was back titrated with ferrous ammonium sulphate until the colour was changed to green and the titrated value was noted down. A blank titration was also carried out.

$$\text{\%Organic carbon} = N\ \frac{(B-C)}{\textit{weight of the soil}}\times 0.003\times 100$$

% organic matter = organic carbon × 1.724 (**Van Bemmelen factor**)

B = Volume of 0.5N ferrous ammonium sulphate required to neutralize 10ml of 1N of $K_2Cr_2O_{7\,i.e}$. blank titration (blank reading).

C = Volume of 0.5N ferrous ammonium sulphate needed for titration of soil sample (reading with soil).

3.4.6. Determination of available nitrogen

The determination of available nitrogen in soil was done with the help of Kjeltech semi auto-analyser which was given by **Subbiah and Asija (1956)**. In a distillation tube 5g of soil sample was transferred and 5ml of distilled water was added in order to wash out the adhering soil particles at the neck of the cylinder. Then to semi auto-analyser distillation tube was setup and in a 250ml conical flask 20ml boric acid of pH 4 was taken and outlet tube was dipped in conical flask containing boric acid. Now button which is mounted on instrument was pressed so 25ml of sodium hydroxide solution was added to distillation tube. The distillation process was set upto 9 min. The main principle behind this process was nitrogen is released in the form of ammonia this ammonia gets trapped by boric acid and forms green colour then the contents are titrated with 0.02N sulphuric acid until it changes to pink colour. A blank should also be carried out.

Materials and methods

$$\text{Available N}\left(\text{kg ha}^{-1}\right)=\frac{(S-V)\times 0.02\times 14\times 2.24\times 10^{6}}{1000\times 5}$$

S = Sample titration reading

V = Blank titration reading

3.4.7. Determination of available phosphorus

Soil phosphorus in the soil samples were estimated by **Olsen's** method as soil samples were having pH more than 5.5. When pH is <5.5 **Bray's** method is followed and when pH is >5.5 Olsen's method is followed

Olsen's method of estimation

Firstly reagent A and reagent B was prepared. In a conical flask 2.5g of soil was taken and to this soil sample 50ml of Olsen's reagent (0.5 M $NaHCO_3$) was added and a pinch of activated charcoal was also added. The contents were shaken for 30min in a shaker and filtered through Whatman no: 1 filter paper. From this filtrate 10ml of aliquot was taken in a 25ml volumetric flask and 2-3 drops of nitro phenol indicator was added drop by drop until the colour was changed to yellow colour. After the colour was changed to yellow 2.5ml of 2.5M H_2SO_4 was added drop by drop till the yellow colour was disappeared. Then 8ml of reagent B (Murphey Riley colour reagent) was added and volume was make up to 25ml with distilled water. After 15min readings were observed in spectrophotometer at 730nm. Standards of 0, 1, 2, 3, 4, 5ppm were also prepared. A blank was also carried out. Calibration curve was plotted using daily prepared standard solutions.

Preparation of reagent A

SOLUTION 1: In 1000ml volumetric flask add 140ml of 2.5M H_2SO_4 and make up the volume with distlled water.

➡ SOLUTION 2: Take 12g of ammonium molybdate chemical in a beaker and add 250ml of distilled water.

➡ SOLUTION 3: Take 0.291g of antimony potassium tartarate chemical in beaker and add 100ml of distilled water. Then take 2lit volumetric flask and mix 1+2+3rd solutions and make up the volume.

Preparation of reagent B (Murphey Riley colour reagent)

step 1: Take 200ml of reagent A in a 250ml beaker

➡ step 2: To this reagent A present in beaker add 1.056g of ascorbic acid and mix throughly

➡ step 3: The ragent B prepared should be always fresh and reagent A can be stored for one month.

$$\text{Available N}\left(\text{kg ha}^{-1}\right)=\frac{\text{R}\times\text{Volume of extractant}\times 2.24\times 10^{6}}{\text{Volume of aliquot}\times 2.24\times\text{wt of soil}}$$

Where R = μg P in the aliquot (obtained from standard curve).

3.4.8. Determination of available potassium

The determination of available potassium in soil was done by use of flame photometer instrument and extracted using neutral normal ammonium acetate solution which was given by **Schollenberger and Simon, 1945.** In 100ml conical flask 5g of soil sample was taken and to that 25ml of extractant was added and left 5min for shaking. After shaking the contents were filtered immediately with Whatman no: 1 filter paper into a beaker. The potassium concentration in the filtrate was estimated by flame photometer. Standards of 50ppm and 100ppm were prepared from 1000ppm potassium chloride solution (KCl). First the standard reading was taken followed by sample reading.

Dilution factor = 25/5 = 5 times

Reading of the flame photometer for the test sample = C

Available Potassium (kg ha^{-1}) = C × 5 × 2.24

3.4.9. Determination of exchangeable calcium

The determination of exchangeable calcium in soil was done by use of neutral normal ammonium acetate solution which was given by **Jackson, 1973.** In 100ml conical flask 5g of soil sample was taken and to that 25ml of extractant was added. Then the contents were shaken in a rotatory shaker for 5min and immediately the contents were filtered with Whatman no: 1 filter paper and calcium was determined after pre-treatment. After pre-treatment 5ml of sample was pipetted out in conical flask and diluted to 25ml with distilled water. Then 5 drops of 4N sodium hydroxide buffer solution and 50mg of Muroxide indicator was added to the sample. The sample was titrated with 0.01N EDTA until the colour was changed from orange red to purple and the titrated value was noted down.

$$\text{Amount of Ca (Meq/lit)} = \frac{\text{R}\times\text{Normality of EDTA}\times 1000}{\text{Aliquot (ml) taken}}$$

$$\text{Amount of Ca (Meq/100g)} = \frac{100}{\textit{Soil weight (g)}}\ \#\ \frac{\textit{extract volume (ml)}}{100}\ \#\ \textit{Ca in meq/lit}$$

R=Volume (ml) of standard EDTA used in titration

3.4.10. Determination of exchangeable calcium +magnesium

The determination of exchangeable calcium + magnesium in soil was done by use of neutral normal ammonium acetate solution which was given by **Jackson, 1973.** In 100ml conical flask 5g of soil sample was taken and to that 25ml of extractant was added. Then the contents were shaken in a rotatory shaker for 5min and immediately the contents were filtered with Whatman no: 1 filter paper and calcium + magnesium were determined after pre-treatment. After pre-treatment

5ml of sample was pipetted out in conical flask and diluted to 25ml with distilled water. Then 10 drops of ammonium chloride-ammonium hydroxide buffer solution and 4 drops of Erichrome black- T indicator was added to the sample. The sample was titrated with 0.01N EDTA until the colour was changed from wine red to blue and the titrated value was noted down. To get the magnesium content in soil sample calcium + magnesium content in soil sample was subtracted with calcium content in soil sample.

$$\text{Amount of Ca + Mg (Meq/lit)} = \frac{\text{R} \times \text{Normality of EDTA} \times 1000}{\text{Aliquot (ml) taken}}$$

$$\text{AmountofCa+Mg(Meq/100g)} = \frac{100}{\textit{Soil weight (g)}} \# \frac{\textit{extract volume (ml)}}{100} \# \textit{Ca in meq/lit}$$

Amount of Mg (Meq/lit) = Ca + Mg (Meq/lit) – Ca (Meq/lit)

R=volume (ml) of standard EDTA used in titration

Pre-treatment to estimate exchangeable calcium and calcium+magnesium in soil

1.Transfer ammonium acetate extractant to 250ml beaker and evaporate the content on water bath.	2.Wash down the walls of the beaker with small quantity of water and again evaporate.	3.Add 1ml of nitric acid and 3ml of hydrochloric acid.	4.Dissolve the residue after evaporation in 20ml 0.1N acetic acid.	5.Filter through Whatman no: 1 filter paper into 50ml volumetric flask .And dilute the volume

3.4.11. Determination of available sulphur

Sulphur was estimated in the soil by **Turbidimetric** method (**Chesnin and Yien, 1950**). 10g of soil sample was taken and 50ml of calcium chloride extractant of 0.15% was added to the samples. The contents were then shaken in a shaker for 30min and filtered through Whatman no: 1 filter paper. Then 10-20ml of filtrate was taken and added in 25ml volumetric flask and 2ml of gum acacia solution was added to stabilize the contents and volume was make up. Then the contents was again shaken in a shaker for 10min and readings were taken on spectrophotometer at 340nm. Working standards of 0.25, 0.5, 1.0, 2.5, and 5.0 were also prepared from 100ppm potassium sulphate solution.

$$\text{Available Sulphur (mg Kg}^{-1}\text{)} = R \times \frac{50}{10} \times \frac{1}{10}$$

Where, R stands for S content in μg as read on X-axis

3.4.12. Determination of soil pH

Soil pH was determined by using digital pH meter described by **Jackson, 1973.** In 50 ml beaker 1g of soil sample was taken and to that 25 ml of distilled water was added in ratio of 1:2.5 and reading were noted down.

Table: 3.4.12a. Classification of soil samples under pH ranges

Classes	Limit
Moderately acidic	5.6 – 6.5
Neutral	6.6 – 7.3
Strongly acidic	<5.5
Alkaline	>7.3

3.4.13. Determination soil electrical Conductivity

Soluble salts in soil was determined by using EC meter described by **Jackson, 1973.** In 50 ml beaker 1g of soil sample was taken and to that 25 ml of distilled water was added in ratio of 1:2.5 and reading were noted down. The readings were in dSm^{-1}.

Table: 3.4.13a. Classification of soil samples under salinity ranges

Classes	EC(dSm^{-1})
All crops	<0.7
Most crops	0.7-2.0
Salt tolerant	2.0-10.0
Most halophytes	10-32
No crops(sea water)	>32

Table: 3.4. Rating chart of different nutrients

Nutrients	LOW	MEDIUM	HIGH
Organic carbon (%)	<0.5	0.5-0.75	>0.75
Available N(kg ha-1)	<280	280-560	>560
Available P (kg ha-1)	<12.5	12.5-25	>25
Available K (kg ha-1)	<135	135-335	>335
Available S (mg kg-1)	<10	10-20	>20
	DEFICIENT	SUFFICIENT	
Magnesium(Meq/100g)	<1.5	>1.5	-
Calcium (Meq/100g)	<1.0	>1.0	-

3.5 Evaluation of Soil Nutrient Status

In order to compare the levels of soil fertility of one area with those of another it was necessary to obtain a single value for each nutrient. Nutrient index (N.I) value is a measure of nutrient supplying capacity of soil to plants. The nutrient index approach introduced by (Parker *et al.* 1951) has been adopted and modified by several researchers and national /international organizations such as ICAR -NBSS&LUP13, Ministry of Agriculture (Govt. of India), FAO etc. This index is used to evaluate the fertility status of soils based on the samples in each of the three classes, i.e., low, medium and high.

Nutrient Index (N.I.) = (NL × 1 + NM × 2 + NH × 3) / NT

Where, NL: Indicates number of samples falling in low class of nutrient status

NM: Indicates number of samples falling in medium class of nutrient status

NH: Indicates number of samples falling in high class of nutrient status

NT: Indicates total number of samples analysed for a given area.

In this percent assessment a nutrient index less than 1.5 denotes low category and that falls between 1.5 and 2.5 represents the medium fertility class. Value of 2.5 and above (max 3.00) signifies a high fertility class in respect of the particular nutrient. The rating of the individual soil test values into the three classes was performed according to the above mentioned limits. (Ramamoorthy and Bajaj, 1969).

Table: 3.5a. Rating Chart of Nutrient Index

Sl.No.	Nutrient Index	Value	Interpretation
1	Low	<1.67	Low fertility Status of the area
2	medium	1.67-2.33	Medium fertility Status of the area
3	high	>2.33	High fertility Status of the area

(Source: Ramamoorthy and Bajaj, 1969)

3.6. Method of water sampling and preservation

First, in this study sample collection of ground water was done in a polythene bottle. Three water samples were collected from eight different sites/villages i.e. a total of 24 ground water samples were collected in Madanapalle Mandalam, Chittor district, Andhra Pradesh. 2-3 drops of toluene was also added in the samples to stop the growth of microbes. **The brief description of water samples were given in Appendix: 6.**

3.7. Physico – Chemical analysis

The water samples collected were analysed for chemical and water quality parameters in laboratory which includes pH, EC, sodium, potassium, bicarbonate, chloride, calcium, magnesium, SAR, RSC, SSP, PI and KR using standard methods as described by APHA, 1992. Methods followed and equipment used are given in Table-3.1b and the range of all parameters range in given in the Table-3.7.

3.7.1. Measurement of water pH

For measurement of water pH suitable amount of water sample was taken in 50ml beaker and measured using pH meter as described by **APHA, 1992.**

3.7.2. Measurement of water salinity

For measurement of water EC suitable amount of water sample was taken in 50ml beaker and measured using EC meter as described by **APHA, 1992** and expressed in dSm-1.

3.7.3. Determination of alkalinity (carbonate and bicarbonate)

Simple **Acidimetric titration method** was used to determine Carbonates and bicarbonates in water samples (**APHA, 1992**). Carbonates and bicarbonates was estimated by using 0.05N sulphuric acid solution. 5ml of water sample was pipetted out into a conical flask and diluted to 25ml with distilled water. Then 2 drops of phenolphthalein indicator was added to the samples the colour did not changed to pink which indicated that there is no presence of carbonates in the water samples. So to the same water samples 2 drops of methyl orange indicator was added the initial colour changed to yellow then it was titrated with 0.01N sulphuric acid until the final colour changed to rose red and the burette volume was noted down.

$$CO_3^{2-} Meq/lit = \frac{Normality\ of\ H_2SO_4 \#\ vol\ of\ H_2SO_4 \#\ 1000}{ml\ of\ aliquot\ taken}$$

$$CO_3^{2-}\ gram/lit = \frac{Normality\ of\ H_2SO_4 \#\ volof\ H_2SO_4}{mlof\ aliquot\ taken} \times Eq.wt\ of\ CO_3^{-}\ (30)$$

$$HCO_3^{-} = \frac{Normality\ of\ H_2SO_4 \#\ vol\ of\ H_2SO_4 \#\ 100}{ml\ of\ aliquot\ taken}$$

$$HCO_3^{-} = \frac{Normality\ of\ H_2SO_4 \#\ volof\ H_2SO_4}{mlof\ aliquot\ taken} \times Eq.wt\ of\ HCO_2^{3-}\ (61)$$

3.7.4. Determination of chloride

Chloride in water samples was determined by **Mohr's titration method (APHA, 1992)**. 5ml of water sample was pipetted out in a conical flask and was diluted with 25ml of distilled water. Then 5-6 drops of potassium chromate indicator (K_2CrO_4) was added to samples and the initial colour was changed to dark yellow. Then it was titrated with 0.02N silver nitrate solution until the final colour was changed to brick red.

$$Cl^{-}\ Meq/lit = \frac{Normality\ of\ AgNO_3 \times vol\ of\ AgNO_3 \times 1000}{ml\ of\ aliquot\ taken}$$

$$Cl^{-}\ gram/lit = \frac{Normality\ of\ AgNO_3 \times vol\ of\ AgNO_3}{ml\ of\ aliquot\ taken} \times Eq.wt\ of\ Cl^{-}\ (35.5)$$

3.7.5. Determination of calcium + magnesium

Calcium and magnesium in water samples were determined by **Complexometric** titration method. 5ml of water sample was taken in a conical flask and was diluted it with 25ml of distilled water. Then 1ml of ammonium hydroxide-ammonium chloride buffer solution was added and a pinch of carbamate crystals were added to prevent the interference with other metal ions. Then 2-3 drops of Erichrome Black-T indicator was added the initial colour was pink. It was titrated with 0.01N EDTA solution until the final colour blue appeared.

$$Ca^{2+}\ and\ Mg^{2+}\ Meg/lit = \frac{Normality\ of\ EDTA \times vol\ of\ EDTA \times 1000}{ml\ of\ aliquot\ taken}$$

$$Ca^{2+} \text{ and } Mg^{2+} \; gram/lit = \frac{Normality \; of \; EDTA \times vol \; of \; EDTA \times 1000}{ml \; of \; aliquot \; taken} \times Eq.wt \; of \; Ca^{2+} \text{ and } Mg^{2+}$$

3.7.6. Determination of calcium

Calcium in water samples was determined by **Complexometric** titration method. 5ml of water sample was taken in a conical flask and was diluted it with 25ml of distilled water. Then 5ml of sodium hydroxide buffer solution was added and a pinch of carbamate crystals were added to prevent the interference with other metal ions. Then 40mg of mixed Muroxide indicator (0.5g Muroxide +40gpotassium sulphate) was added the initial colour was changed to pink. It was titrated with 0.01N EDTA solution until the final colour purple appeared.

$$Ca^{2+} \text{ and } Mg^{2+} \; Meg/lit = \frac{Normality \; of \; EDTA \times vol \; of \; EDTA \times 1000}{ml \; of \; aliquot \; taken}$$

$$Ca^{2+} \text{ and } Mg^{2+} \; gram/lit = \frac{Normality \; of \; EDTA \times vol \; of \; EDTA \times 1000}{ml \; of \; aliquot \; taken} \times Eq.wt \; of \; Ca^{2+} \text{ and } Mg^{2+}$$

3.7.7. Determination of potassium

The potassium concentration in water samples were determined by flame photometer as suggested by **APHA, 1992**. 1000ppm of KCl standard stock solutions were prepared from that stock solution 10ml of concentrate was taken and added in 100ml volumetric flask and volume was make up which make 100ppm concentration and also 5ml from 1000ppm stock solution was taken and added in 100 ml volumetric flask and volume was make up which make 50ppm concentration. Standards of 0, 2, 4, 6, 8, 10ppm from 100ppm were prepared in 100ml volumetric flask. The flame photometer was calibrated with standards and readings of samples were taken.

3.7.8. Determination of sodium

The sodium concentration in water samples were determined by flame photometer as suggested by **APHA, 1992**. 1000ppm of NaCl standard stock solutions were prepared from that stock solution 10ml of concentrate was taken and added in 100ml volumetric flask and volume was make up which make 100ppm concentration and also 5ml from 1000ppm stock solution was taken and added in 100 ml volumetric flask and volume was make up which make 50ppm concentration. Standards of 0, 2, 4, 6, 8, 10ppm from 100ppm were prepared in 100ml volumetric flask. The flame photometer was calibrated with standards and readings of samples were taken.

Table: 3.7. Rating chart of water parameters.

Parameters	Suitable	Moderately suitable	Not suitable
pH	6.5 – 8.4	0 – 5.0	>9.5
EC(dSm^{-1})	<0.3	3.0 – 7.0	>7.0
HCO_3^-(Meq/lit)	>1.25	1.25 – 8.5	8.5
Cl^-(Meq/lit)	<4.0	4.0 – 10.0	>10.0
Na^+(Meq/lit)	<3.0	3.0 – 9.0	>9.0

(Source: FAO.org)

3.8. Irrigation Water Quality Index (IWQI)

The water quality indices were derived from primary parameters which refers to its suitability for its use in agriculture. The concentration and composition of dissolved components determine the quality for irrigation. Quality of water plays an important role in determining the salinity or alkalinity of the irrigated area. The requirements for irrigation water quality could differ from one field to the other depending on the cultivated crop pattern as well as the regional soil and climatological conditions *(Babiker et al. 2007).*

IWQI model was applied on the data. This model developed by *(Meireles et.al. 2010)*. In the first step, selection of parameters in this we have to study the Indian standard *(BIS 1991)* for drinking water specification. Here, the physico-chemical parameters along with desirable limits are given. A parameter has to be selected based on its impact in the overall quality of water.

Computation of WQI: The WQI is computed following the three steps

First step: Assigning of weight (w_i) to the selected water parameters (e.g. pH, HCO_3, Cl, EC, Na, K.....etc) according to their relative importance in the overall quality of water.

Second step: Computation of relative weight (W_i) of the chemical parameter using the following equation

$$W_i = w_i / \Sigma\, w_i \text{ (i=1 to n)}$$

Where, $W_{i\,=}$ is the relative weight

w_i = Weight of each parameter

n = Number of parameters

Third step: Assigning of a quality rating scale (q_i) for each parameter, as below:

$$q_i = (C_i / S_i) \times 100$$

Where, $q_{i\,=}$ quality rating

C_i = Concentration of each chemical parameter in each water sample in mg/l

S_i = Guide line value given in BIS 1991

For computation of WQI, the sub-index (SI) is first determined for each chemical parameter, given below:

$$Sl_i = W_i \times q_i$$

$$WQI = \Sigma\, Sl_i \text{ 1-n}$$

Where, Sl_i = sub index of ith parameter

W_i = Relative weight of ith parameter

q_i = Rating based on concentration of ith parameter

n = Number of chemical parameters

Table: 3.8a.The range of WQI, quality status and conceivable usage of water

WQI	Water quality status	Usage
<50	Excellent	Irrigation, industrial and drinking
50-100	Good	Irrigation, industrial and drinking

WQI	Water quality status	Usage
100-200	Poor	Irrigation and industrial
200-300	Very poor	Irrigation
>300	Unsuitable	Appropriate treatment is needed

(*Source: Anbazaghan, 2014*)

Table: 3.8b. Water quality standards (All are in mg/l except pH and EC)

Parameters	Standards
pH	7.5
Electrical conductivity	0.3
Bicarbonate	300
Calcium	75
Magnesium	30
Chloride	250
Potassium	20
Sodium	200

(*Source: BIS 1991*)

3.8.1. Sodium Adsorption Ratio (SAR):

SAR is a measure of the amount of sodium (Na^+) relative to calcium (Ca^{2+}) and magnesium (Mg^{2+}) in the water. It the ratio of sodium concentration divided by square root of one-half of the Ca+Mg concentration. **Richards** calculated the sodium adsorption ratio by the following formula it is expressed in Meq L^{-1}.

$$SAR = \frac{Na}{\sqrt{Ca + Mg / 2}}$$

Table: 3.8.1a. Classification of irrigation water based on SAR (Richards)

SAR	Class
0-10	Low sodium
10-18	Medium
18-26	High
>29	Very high

3.8.2. Soluble sodium percentage:

Soluble sodium percentage gives a clear idea about sodium content which is important for studying sodium hazard. High SSP hinders the growth of the plants and reacts with soil to reduce its permeability. The SSP is calculated by using the following formula it is expressed in Meq L^{-1} (**Todd 1980**).

$[100\times (Na^+ + K^{+)}] / (Ca^{2+} + Mg^{2+} Na^+ + K^+)$

Table: 3.8.2a. Classification of irrigation water based on SSP (Todd)

SSP	Class
<20	Excellent
20-40	Good
40-60	Permissible
60-80	Doubtful
>80	Unsuitable

3.8.3. Permeability Index (PI):

The permeability is affected by consistent use of irrigation water which increases the presence of sodium, calcium, magnesium and bicarbonate in the soil (**Chandu et al. 1995**). The permeability index (PI) is used to measure the suitability of water for irrigation purpose when compared with the total ions in Meq L^{-1}.

$$PI = \frac{Na + \sqrt{HCO_3}}{Ca + Mg + Na} \times 100$$

Table: 3.8.3a. Classification of irrigation water based on PI

PI	Class
>75	Suitable
25-75	Moderately suitable
<25	Not suitable

3.8.4. Kelly's Ratio (KR):

Suitability of water for irrigation purposes is also assessed on the bases of Kelly's ratio (**Kelly 1951**). Ratio of sodium verses calcium and of sodium verses magnesium is used as Kelly's ratios. Groundwater having Kelly's ratio more than one is considered not-suitable for irrigation purposes and ground water having Kelly's ratio less than one is considered suitable for irrigation. Kelly's ratio was calculated by using the following expression it is expressed in Meq L^{-1}.

$$\frac{100}{\text{Soil weight (g)}} \# \frac{\text{extract volume (ml)}}{100} \# \text{Ca in meq/lit}$$

3.8.5. Residual sodium carbonate (RSC):

Residual sodium bicarbonate is an index which is used to measure the alkalinity hazard in water. When sodium concentrated water is applied to clay soil it swells and causes dispersion of soil particles which reduces the infiltration capacity of the soil. The RSC is calculated by using the following formula given by **Eaton (1950)** it is expressed in Meq L^{-1}.

$$\text{RSC (Meq/lit)} = (CO_3^{2-} + HCO3^{-}) - (Ca^{2+} + Mg^{2+})$$

Table: 3.8.5a. Classification of irrigation water based on RSC

RSC	Class
<1.25	Suitable
1.25-2.50	Moderately suitable
>2.50	Not suitable

3.9. Statistical Analysis

The statistical data i.e. mean, standard deviation coefficient of variation, range and correlation coefficient for soil as well as water parameters were computed using SPSS statistical software. The formulas for computing the statistical data are given under:

1. Mean = $\frac{\textit{Sum to all Samples}}{\textit{Total Number of Samples}}$

2. Range = largest value (L) – smallest value (S)

3. Coefficient of Variation = $\frac{\text{Standerd Deviation}}{Mean} \times 100$

4. Standard Deviation = $\sqrt{\frac{RX^2 - \frac{[RX]^2}{n}}{n-1}}$

Where, Σx^2 = Sum of all samples

$(\Sigma x)^2$ = Sum of squares of all samples

n = Total number of samples

5. Coefficient of Correlation **(Pearson)** = $\frac{SS(XY)}{\sqrt{[SS(X)][SS(Y)]}}$

Where, SS (XY) = Sum of the product of x and y

SS(X) = Sum of squares of x

SS(Y) = Sum of squares of y

4

CHAPTER

Results and Discussion

This chapter deals with the results and discussion of soil and water analysis which were collected from the Madanapalle block. 24 water samples and soil samples were collected from different fields and analysed for chemical and physical parameters. The results were correlated, interpreted and possible reasons for variation were discussed under this chapter. The results of each parameter are presented under the following headings.

4.1 Analysis of soil parameters

4.1.1 Physico-chemical properties of soil.

4.1.2 Status of available primary macronutrients in the soil.

4.1.3 Status of available secondary macronutrients in the soil.

4.1.4 Soil nutrient index value.

4.1.5 Correlation matrix between physico-chemical properties of soil.

4.2 Analysis of water parameters

4.2.1 The physico-chemical parameters of ground water

4.2.2 Anions status in ground water.

4.2.3 Cations status in ground water.

4.2.4 Irrigation water quality index.

4.2.5 Correlation matrix between physic-chemical properties of ground water.

4.1. Analysis of soil parameters

4.1.1 Physico-Chemical properties of soil.

The data on the physico-chemical properties of soil are given in the Table-4.1.1 and in Appendix 2 and 3.

Values of water holding capacity in soil samples ranged from 32.5 to 60.6% with 47.1 as a mean value. The standard deviation of water holding capacity is 7.81 and coefficient of variation is 16.5%. The lowest water holding capacity was observed in site 2 sample no: 6 Ponnugutupalem village this is due to low organic carbon content in the soil sample no:6 and highest water holding capacity was reported in site 6 sample no: 16 Ramacharlapallli village as there is high organic carbon content of 0.67% in sample no:16. So the results shows that increase in the organic carbon content higher will be the water holding capacity of the soil. Similar results were reported by Deb *et al.* (2013) in south Sikkim. Values of bulk density in soil samples ranged from 1.16- 1.43 g/cm^3 with 1.30 as a mean value. The standard deviation of bulk density is 0.06 and coefficient of variation is 5.27%. Site 7 sample no: 20 Ganganavaripalli village reported lowest bulk density and site 6 sample no: 16, site 3 sample no: 7 Ramacharlapallli and Nayabandapalli village reported highest bulk density. The highest bulk density was reported in cultivated land due to high intensive agricultural practises in the field which resulted in soil compaction. Similar reports were reported by Chaudhari *et al.* (2013) in Coimbatore district. Values of particle density in soil samples ranged from 2.08 to 2.67 g/cm^3 with 2.37 as a mean value. The standard deviation of particle density is 0.12 and coefficient of variation is 5.28%. Site 1 sample no: 1 Chipilli village reported lowest particle density this may be due to presence of high organic carbon content in the soil sample no: 1 whereas site 5 sample no: 14 Basinikonda village reported highest particle density. Values of porosity in soil samples ranged from 38.3 to 51.8% with 44.98 as a mean value. The standard deviation of porosity is 3.77 and coefficient of variation is 8.39%. The lowest porosity was observed in site 2 sample no: 5 Ponnugutupalem village and highest porosity was reported in site 7 sample no: 20. Ganganavaripalli village. When soil particles get eroded they fill up the pore spaces there by increasing the bulk density of the soil and decreasing the porosity the same was reported in the results. Sample no: 5 showed low porosity due to high bulk density (1.39g/cm^3) of 5th sample and sample no: 20 showed high porosity due to low bulk density (1.16g/cm^3) of 20th sample.

Table: 4.1.1. Statistical analysed data on physico-chemical parameters of soil

Soil parameter	Mean	Range	S.D±	C.V (%)
Bulk density (Mg/m$^{3)}$	1.30	1.16-1.43	0.06	5.27%
Particle density (Mg/m^3)	2.37	2.08-2.67	0.12	5.28%
Water holding capacity(%)	47.1	32.5-60.6	7.81	16.5%
Porosity (%)	44.9	38.3-51.8	3.77	8.39%
pH	7.54	5.87-8.50	0.62	8.24
EC(dSm^{-1})	0.45	0.072-1.264	0.32	72.7
Organic carbon (%)	0.59	0.24-0.95	0.20	34.5

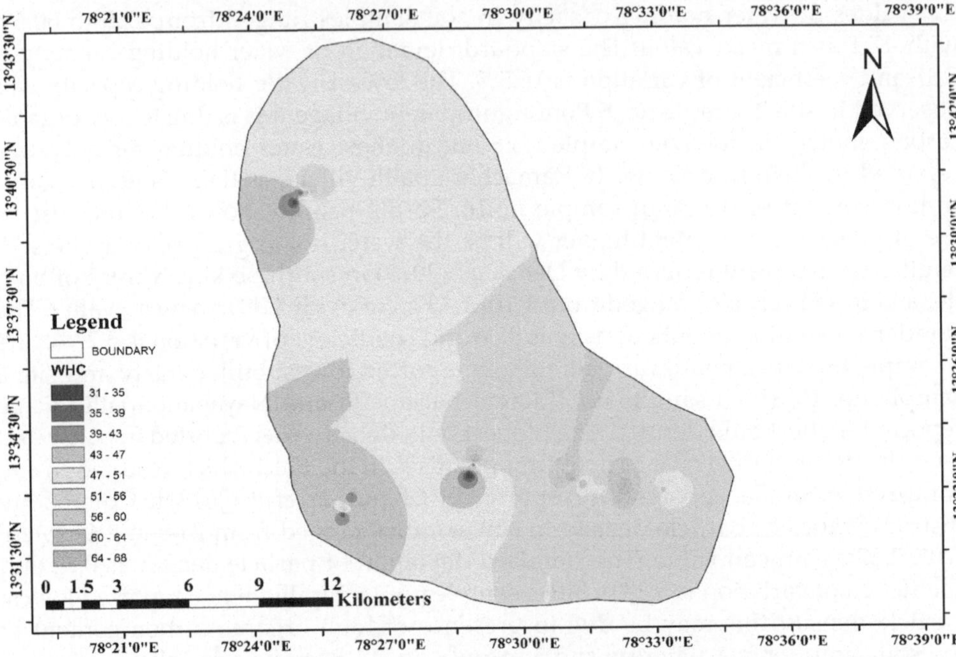

Fig. 4.1.1a: Spatial variability map of Water Holding Capacity of different villages in Madanapalle block, Chittor district, Andhra Pradesh.

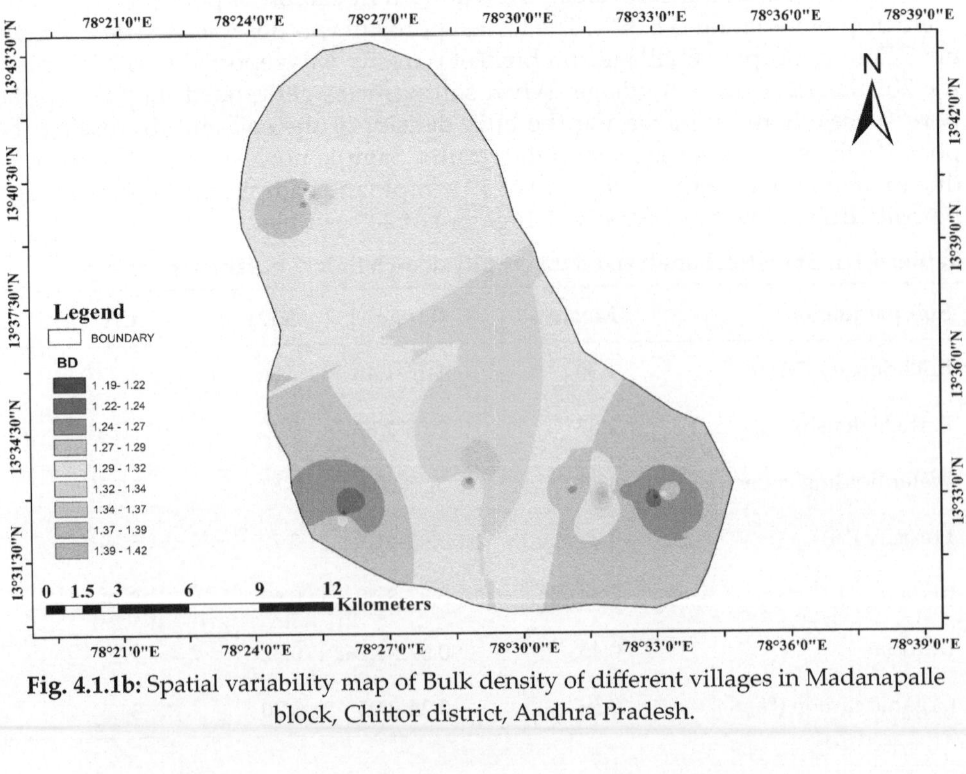

Fig. 4.1.1b: Spatial variability map of Bulk density of different villages in Madanapalle block, Chittor district, Andhra Pradesh.

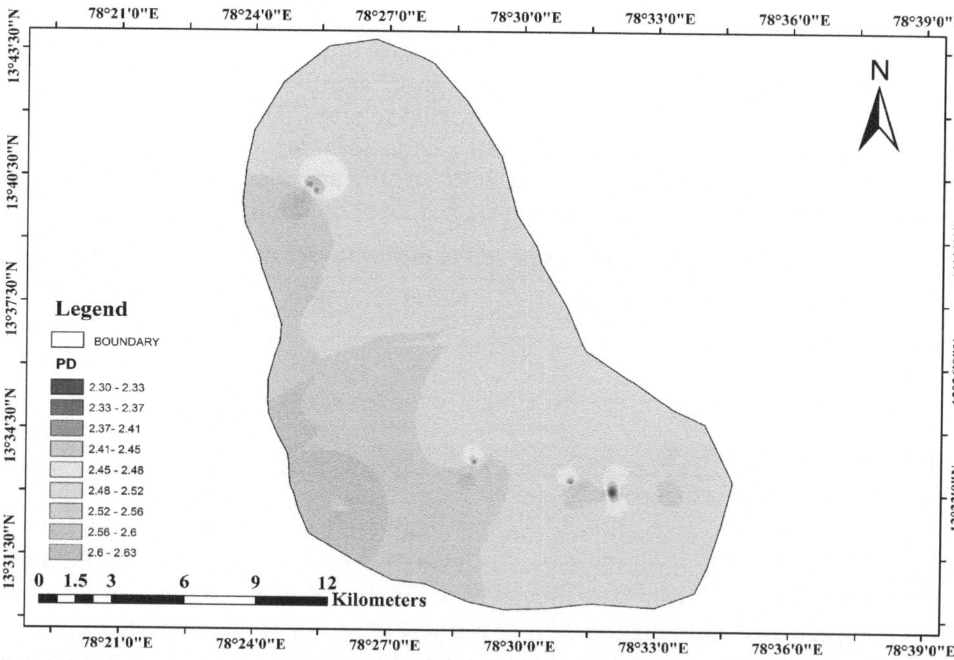

Fig. 4.1.1c: Spatial variability map of Particle density of different villages in Madanapalle block, Chittor district, Andhra Pradesh.

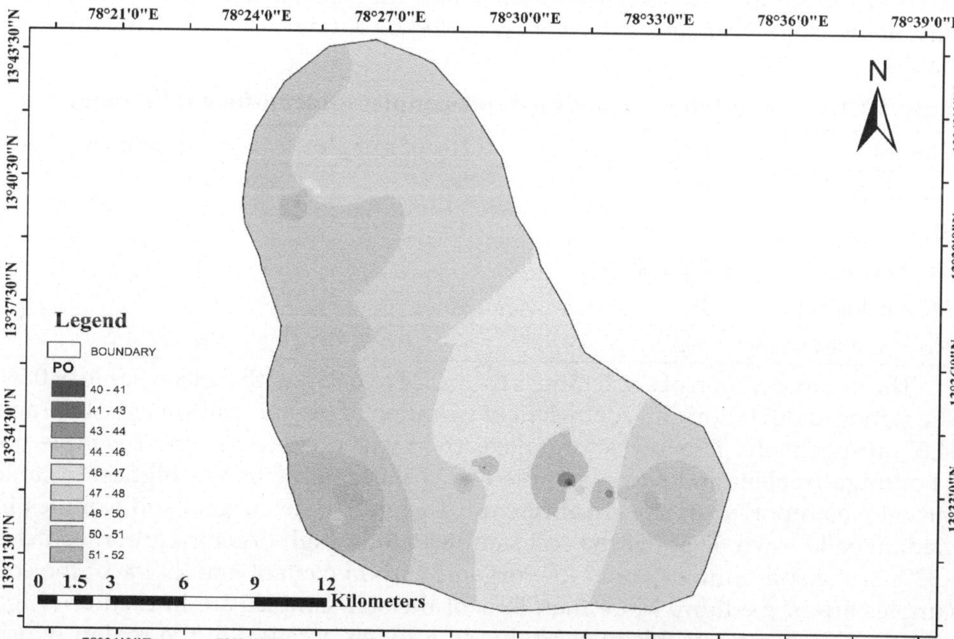

Fig. 4.1.1d: Spatial variability map of Porosity of different villages in Madanapalle block, Chittor district, Andhra Pradesh.

Values of pH in a soil samples ranges from 5.87 to 8.50 with 7.54 as a mean value. The standard deviation of soil pH is 0.62 and coefficient of variation is 8.24%. The lowest pH value was observed in site: 7 sample no: 21 Ganganavaripalli village whereas highest pH value was reported in site: 8 sample no: 22 Barlapalli village. The results concluded that 62.5% of the samples are alkaline in nature, 29.16% of the samples are neutral and 8.32% of the samples are acidic in nature. Similar results were reported by Kumar *et al.* (2011) in Tamil Nadu state.

Table: 4.1.1a. Classification of analysed soil samples under different pH range

Classes	Limit	No. of samples	% of samples
Moderately acidic	5.6-6.5	1	4.16%
Neutral	6.6-7.3	7	29.1%
Strongly acidic	<5.5	1	4.16%
Alkaline	>7.3	15	62.5

(Source: Natural Resource Conservation service)

Values of EC in a soil samples ranged from 0.072 to 1.264 DSm^{-1} with 0.45 as a mean value. The standard deviation and coefficient of variation of soil EC were 0.32 and 72.2% respectively. The lowest EC content was observed in site: 2 sample no: 6 Ponnugutupalem village whereas highest EC content was reported in site: 7 sample no: 19 Ganganavaripalli village. According to the results it is shown that 83.3% of the samples are in permissible range suitable for all type of crops and 16.6% of the samples are slightly higher than the permissible range suitable for most of the crops. Similar results were reported by Kumar *et al.* (2017) in Himachal Pradesh.

Table: 4.1.1b. Classification of analysed soil samples under different EC range

Classes	Limit	No. of samples	% of samples
All crops	<0.7	20	83.3%
Most crops	0.7-2.0	4	16.6%
Salt tolerant	2.0-10.0	-	0%
Most halophytes	10-32	-	0%
No crops(sea water)	>32	-	0%

The organic carbon of a soil ranges from 0.24 to 0.95%with a mean value of 0.59. The standard deviation and coefficient of variation of organic carbon were 0.20 and 34.5% respectively. The lowest organic carbon was observed in site 2 sample no: 6 Ponnugutupalem and site: 8 sample no: 22 Barlapalli whereas highest organic carbon was reported in site: 1 sample no: 1 Chipilli. Out of total soil samples in Madanapalle town 62.5% of the soil samples are in high organic carbon content, 8.33% of the soil samples are in low organic carbon content and 29.1% of the soil samples are in medium. More than 50% of the soil samples are in medium-high organic carbon content this may be due to low temperature in the region which suppress the microbial and enzymatic activity thereby less decomposition of the organic matter and accumulation on the surface of the soil.

Table: 4.1.1c. Classification of soil samples under different organic carbon range

Classes	Limit	No. of samples	% of samples
Low	<0.25	2	8.33%
Medium	0.25 – 0.50	7	29.1%
High	>0.50	15	62.5%

(Source: Muhr et.al, 1963)

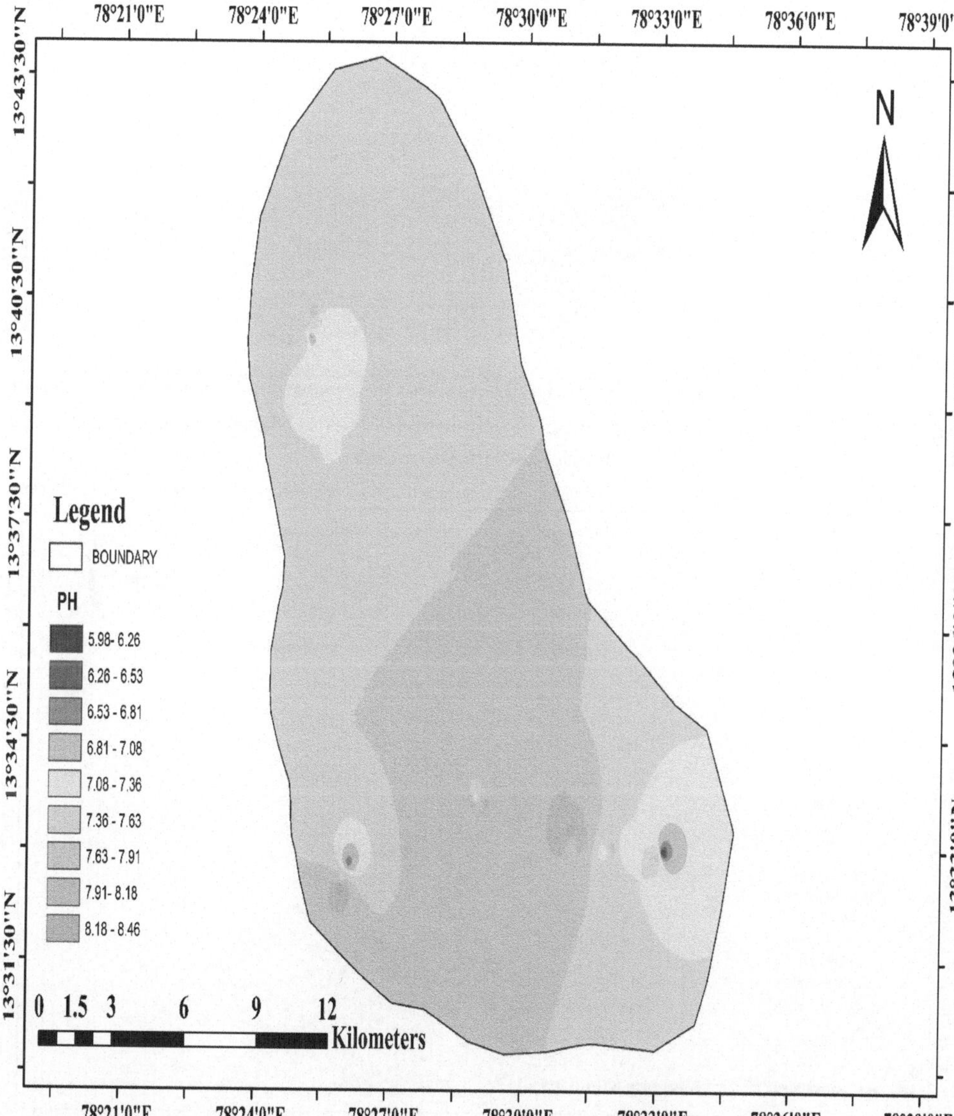

Fig. 4.1.1e: Spatial variability map of pH of different villages in Madanapalle block, Chittor district, Andhra Pradesh.

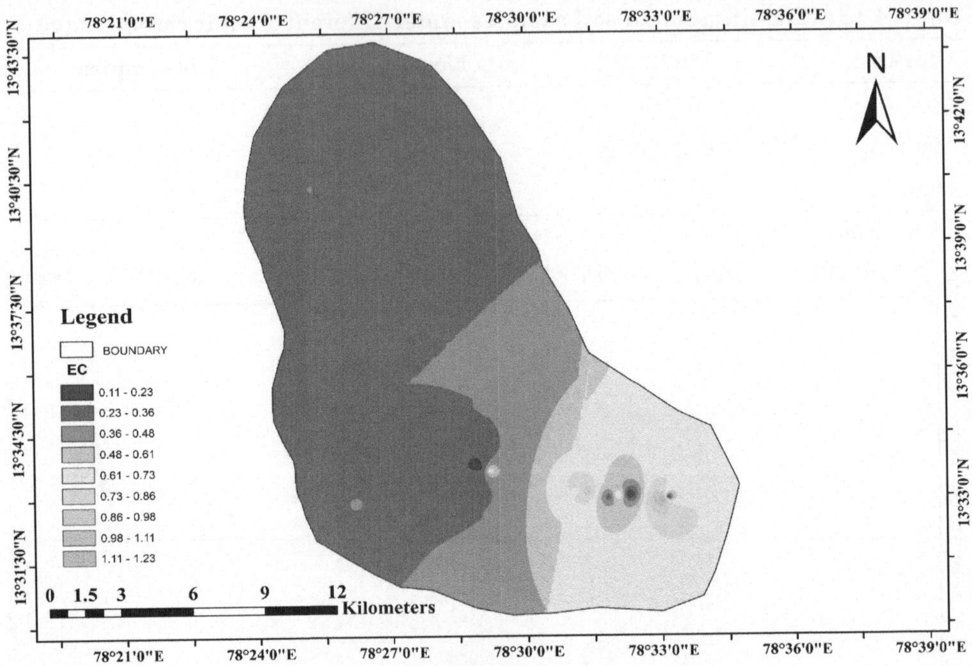

Fig. 4.1.1f: Spatial variability map of EC of different villages in Madanapalle block, Chittor district, Andhra Pradesh.

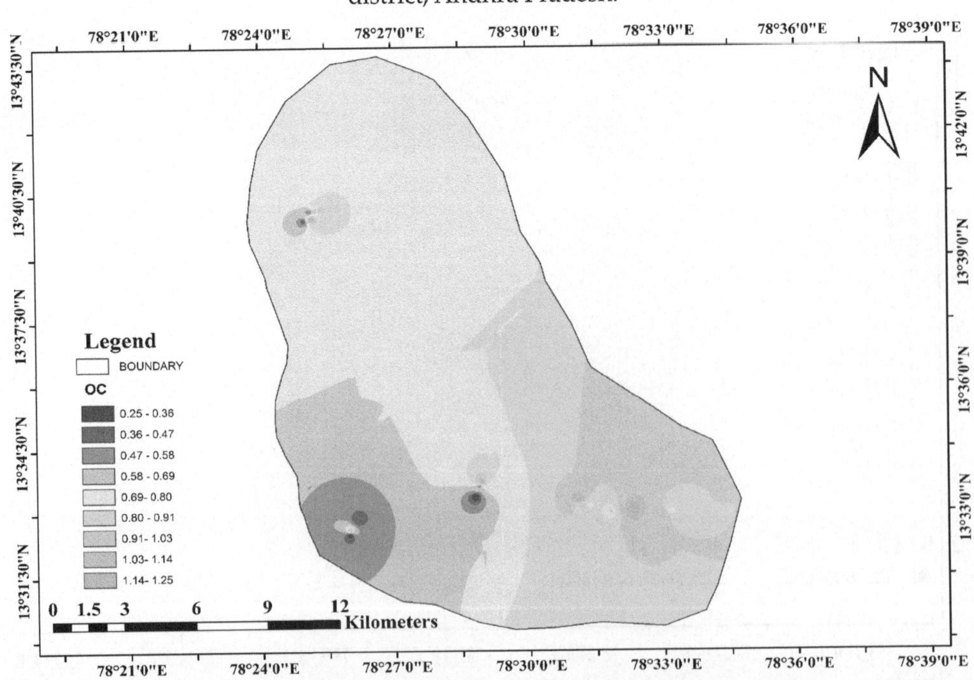

Fig. 4.1.1g: Spatial variability map of organic carbon of different villages in Madanapalle block, Chittor district, Andhra Pradesh.

4.1.2. Status of available primary macronutrients in the soil

The data on the primary macronutrients of soil is given in the Table-4.1.2 and in Appendix 4.

Values of nitrogen in soil samples ranged from 72.31 to 765.9 kg ha^{-1} with 248.6 as a mean value. The standard deviation and coefficient of variation of nitrogen were 204.8 and 82.3% respectively. The lowest nitrogen content was observed in site: 8 sample no: 22 Barlapalli whereas highest nitrogen content was reported in site: 2 sample no: 5 Ponnugutupalem. 70.83% of the samples showed low levels of nitrogen, 16.6% of the samples showed medium levels and 12.5% of the samples showed high levels. Similar results were observed with Dameshwar *et al.* (2018) in Matya village of Kasdol block, Chhattisgarh.

Table: 4.1.2a. Classification of soil samples under different nitrogen range

Classes	Limit	No. of samples	% of samples
Low	<280	17	70.8%
Medium	280-560	4	16.6%
High	>560	3	12.5%

(Source: Muhr et.al, 1963)

Values of phosphorus in soil samples ranged from 1.25 to 31.2 kg ha^{-1} with 13.7 as a mean value. The standard deviation and coefficient of variation of phosphorus were 9.77 and 71.2% respectively. The lowest phosphorus content was observed in site: 2 sample no: 4 Ponnugutupalem whereas highest phosphorus content was reported in site: 5 sample no: 15 Basinikonda. Out of total soil samples 54.1% of the samples are in low phosphorus range this is due to if we look at the pH of soil samples 62.5% of the samples are in alkaline in nature which caused the higher fixation of phosphorus ion by Ca, Mg and Na oxides whereas, 29.1% of the samples are in medium range and 16.6% are in high range.

Table: 4.1.2b. Classification of soil samples under different phosphorus range

Classes	Limit	No. of samples	% of samples
Low	<12.5	13	54.1%
Medium	12.5-25	7	29.1%
High	>25	4	16.6%

(Source: Ramamoorthy and Bajaj, 1969)

Values of potassium in soil samples ranged from 89.6 to 1448.8 kg ha^{-1} with 481.7 as a mean value. The standard deviation and coefficient of variation of potassium were 308.6 and 64.0% respectively. The lowest potassium content was observed in site: 7 sample no: 21 Ganganavaripalli village and site: 6 sample no: 16 Ramacharlapallli village whereas highest potassium content was reported in site: 7 sample no: 19 Ganganavaripalli. Out of total soil samples 66.66% of the samples are in high potassium range the reason may be due to presence of elite rich potassium minerals in these soils, 25% of the soil samples are in medium range and 8.33% of the soil samples are in low potassium range. Similar results were observed with Jain *et al.* (2014) who studied on soil fertility status of Lunawada Taluk of Gujarat.

Table: 4.1.2c. Classification of soil samples under different potassium range

Classes	Limit	No. of samples	% of samples
Low	<135	2	8.3%
Medium	135-335	6	25%
High	>335	16	66.6%

(Source: Ramamoorthy and Bajaj, 1969)

Table: 4.1.2. Statistical data on primary macronutrients of soil

Soil parameters	Range	Mean	SD±	C.V (%)
Nitrogen	72.3-765.9	248.6	204.8	82.3
Phosphorus	1.25-31.2	13.7	9.77	71.2
Potassium	89.6-1448.8	481.7	308.6	64.0

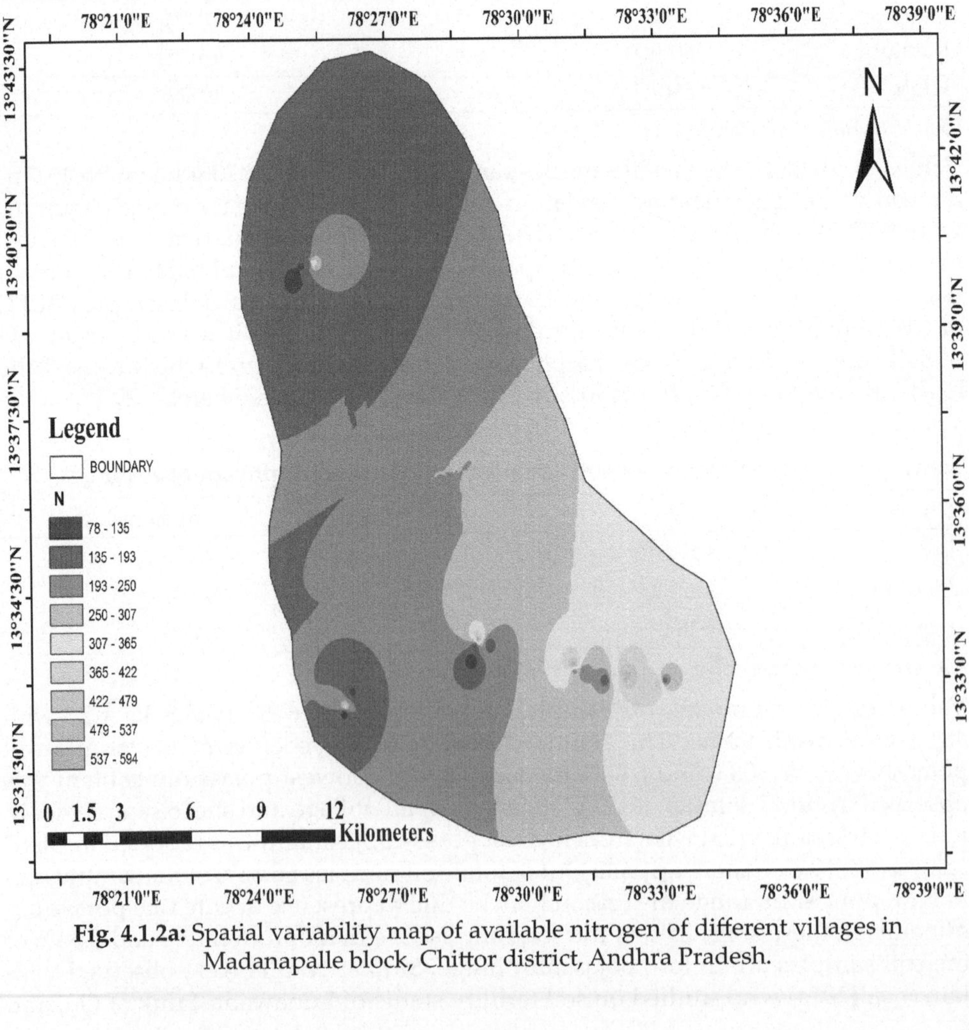

Fig. 4.1.2a: Spatial variability map of available nitrogen of different villages in Madanapalle block, Chittor district, Andhra Pradesh.

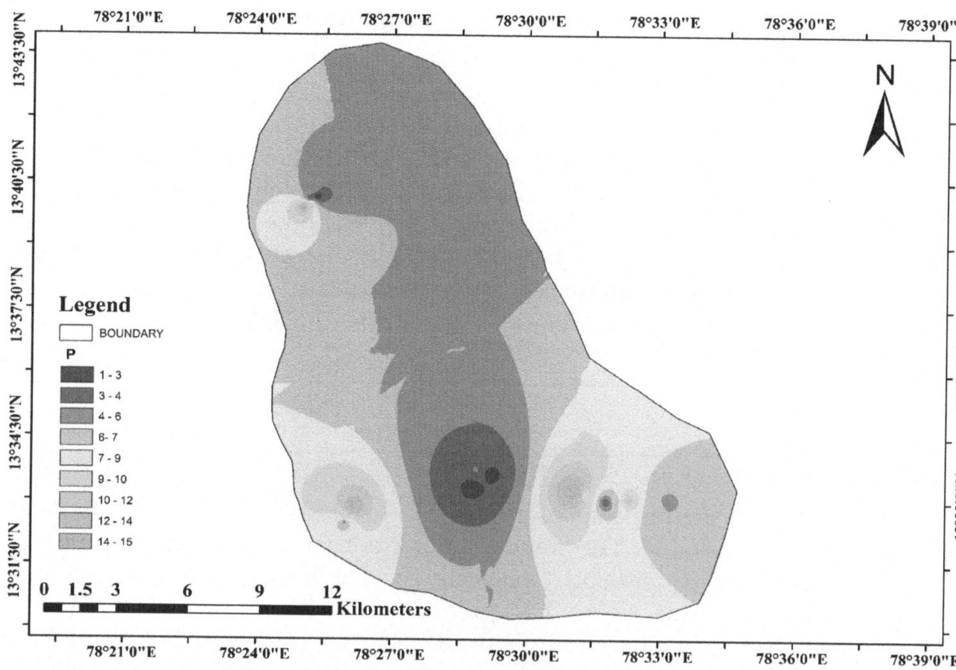

Fig.: 4.1.2b: Spatial variability map of available phosphorus of different villages in Madanapalle block, Chittor district, Andhra Pradesh.

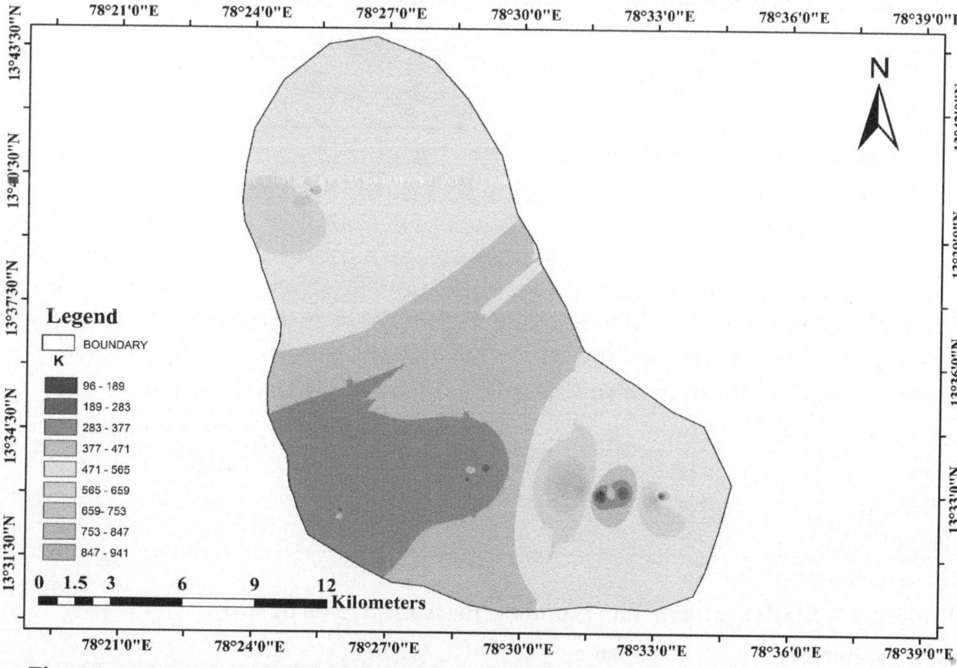

Fig. 4.1.2c: Spatial variability map of exchangeable potassium of different villages in Madanapalle block, Chittor district, Andhra Pradesh.

4.1.3. Status of available secondary macronutrients in the soil

The analysed data on the secondary macronutrients of soil is given in the Table-4.1.3 and in Appendix 5.

Values of calcium in soil samples ranged from 0.40 to 9.08 Meq/100g with 2.96 as a mean value. The standard deviation and coefficient of variation of calcium were 12.34 and 79.1% respectively. The lowest calcium content was observed in site: 7 sample no: 21 Ganganavaripalli village whereas highest calcium content was reported in site: 5 sample no: 15 Basinikonda. 37.5% of the soil samples showed low levels and 62.5% of soil samples are in high range.

Table: 4.1.3a. Classification of soil samples under different calcium range

Classes	Limit	No. of samples	% of samples
Low	<1.5	9	37.5%
High	>1.5	15	62.5%

(Source: Ramamoorthy and Bajaj, 1969)

Values of magnesium in soil samples ranged from 0.04 to 3.70 Meq/100g with 1.05 as a mean value. The standard deviation and coefficient of variation of magnesium were 0.89 and 85.1% respectively. The lowest magnesium content was observed in site: 3 sample no: 7 Nayabandapalli whereas highest magnesium content was reported in site: 7 sample no: 19 Ganganavaripalli. Out of total soil samples 58.3% of the samples are in low range whereas 41.6% of the samples are in high range.

Table: 4.1.3b. Classification of soil samples under different magnesium range

Classes	Limit	No. of samples	% of samples
Low	<1.0	14	58.3%
High	>1.0	10	41.6%

(Source: Ramamoorthy and Bajaj, 1969)

Sulphur content of a soil ranges from 0.14 to 0.45 Mg kg^{-1} with a mean value of 0.25. The standard deviation and coefficient of variation of sulphur were 0.08 and 32.6% respectively. The lowest sulphur content was observed in site: 4 sample no: 12 Kandlamadugu whereas highest sulphur content was reported in site: 3 sample no: 8 Nayabandapalli. In Madanapalle region there is a lesser concentration of sulphur and results reported that all 24 soil samples are deficient in sulphur.

Table: 4.1.3c. Classification of soil samples under different sulphur range

Classes	Limit	No. of samples	% of samples
Low	<10	24	100%
Medium	10-20	0	0%
High	>20	0	0%

(Source: Kanwar, 1976)

Table: 4.1.3. Statistical data on secondary macronutrients of soil

Soil parameters	Range	Mean	SD±	C.V (%)
Calcium	0.40-9.08	2.96	2.34	79.1
Magnesium	0.04-3.70	1.05	0.89	85.1
Sulphur	0.14-0.45	0.25	0.08	32.6

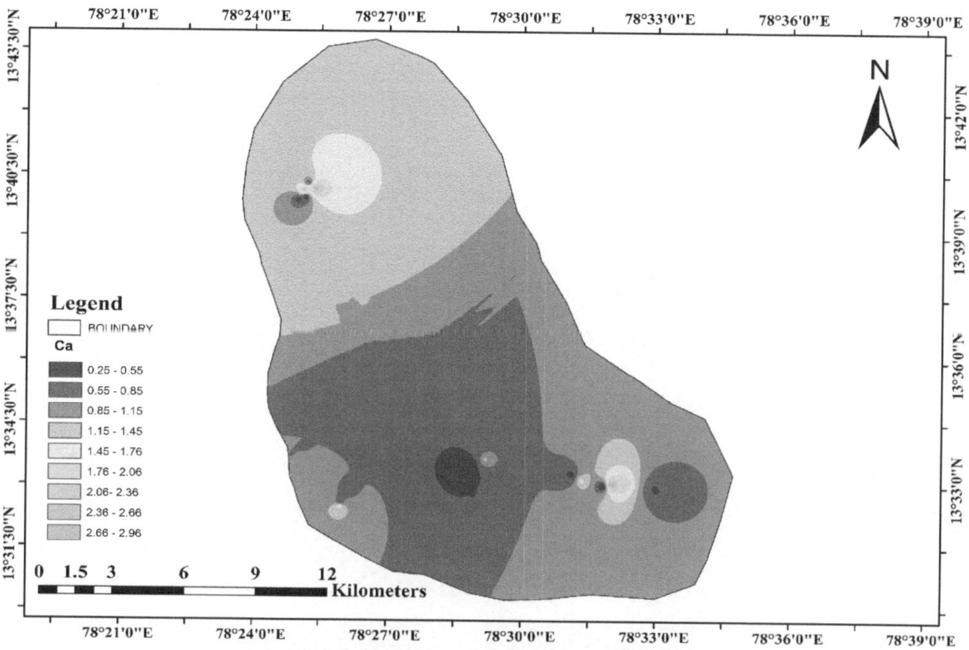

Fig. 4.1.3a: Spatial variability map of exchangeable calcium of different villages in Madanapalle block, Chittor district, Andhra Pradesh.

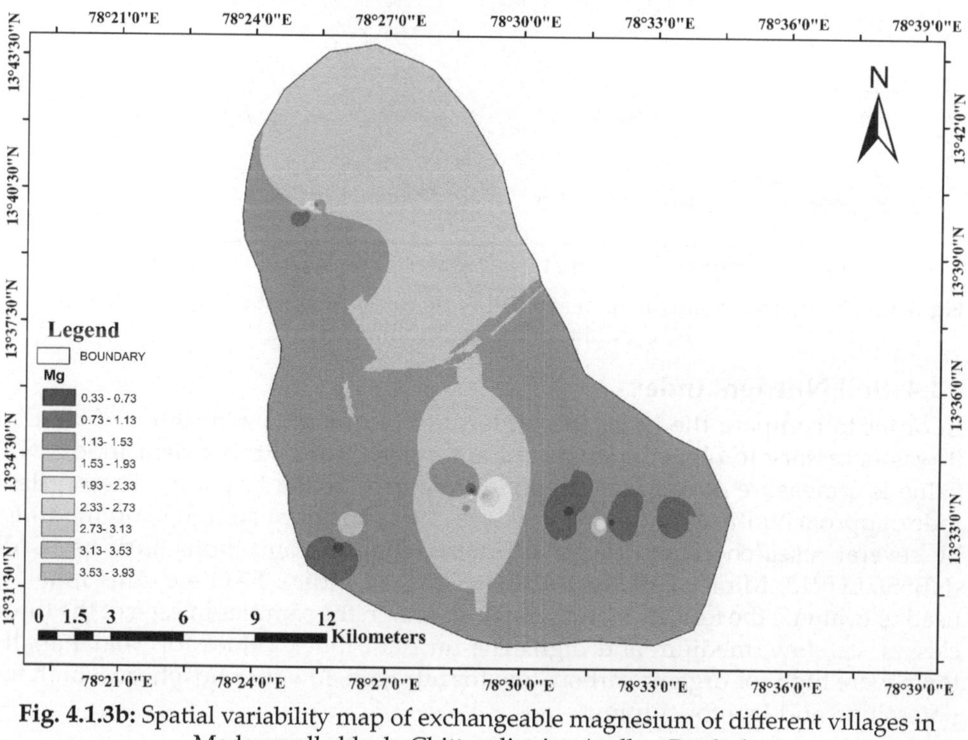

Fig. 4.1.3b: Spatial variability map of exchangeable magnesium of different villages in Madanapalle block, Chittor district, Andhra Pradesh.

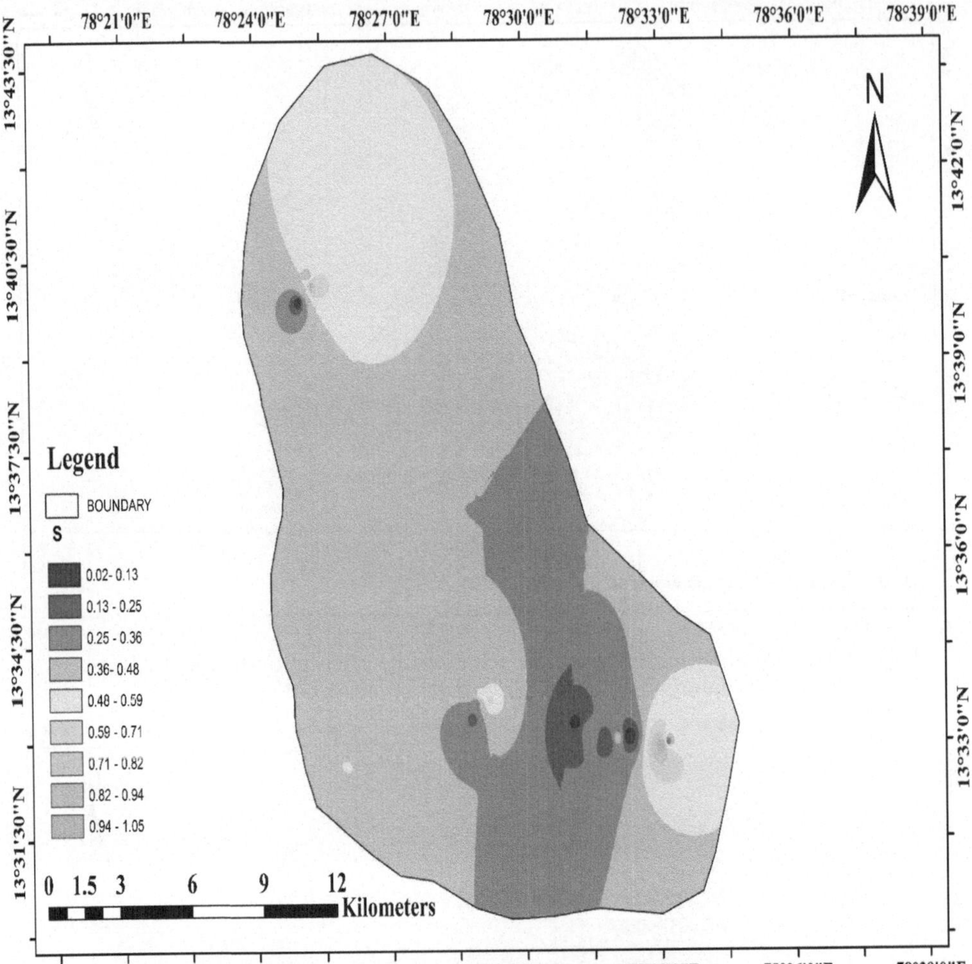

Fig. 4.1.3c: Spatial variability map of available sulphur of different villages in Madanapalle block, Chittor district, Andhra Pradesh.

4.1.4. Soil Nutrient index

In order to compare the levels of soil fertility of one area with those of another it was necessary to obtain a single value for each nutrient. Nutrient index (N.I) value is a measure of nutrient supplying capacity of soil to plants. The nutrient index approach introduced by (Parker *et al.* 1951) has been adopted and modified by several researchers and national /international organizations such as ICAR -NBSS&LUP13, Ministry of Agriculture (Govt. of India), FAO etc. This index is used to evaluate the fertility status of soils based on the samples in each of the three classes, i.e., low, medium and high. The nutrient index values for Madanapalle town were high for organic carbon, low for nitrogen, low for phosphorus, high for potassium, and low for sulphur.

Table: 4.1.4. Nutrient Index values of Madanapalle town in Chittor district of Andhra Pradesh.

Sl.No	Available nutrient	Nutrient index values	Category
1	Nitrogen	1.41	Low
2	Phosphorus	1.62	Low
3	Potassium	2.58	High
4	Sulphur	1	Low
5	Organic carbon	2.54	High

4.1.5. Correlation matrix between physico-chemical properties of soil of different villages in Madanapalle town of Chittor district, Andhra Pradesh.

The data on correlation matrix between physico-chemical properties of soil of different villages in Madanapalle town of Chittor district, Andhra Pradesh were presented in Table: 4.1.5.

The bulk density of the soil is negatively non-significantly correlated water holding capacity (r = -0.282), pH (r = -0.09) organic carbon (r = -0.106), nitrogen (r = -0.11), potassium (r = -0.218), calcium (r = -0.204), magnesium (r = -0.210), sulphur (r = -0.096), negatively significantly correlated with porosity (r = -0.664), EC (r = -0.403), phosphorus (r = -0.539) and positively non significantly correlated with particle density (r = 0.128).

The particle density of the soil is negatively non significantly correlated with organic carbon (r = -0.035), potassium (r = -0.043), sulphur (r = -0.096), positively significantly correlated with porosity(r = 0.652), positively non significantly correlated with pH(r = 0.147), water holding capacity (r = 0.036) and phosphorus(r = 0.139), EC(r =0.228), nitrogen(r =0.224), calcium(r =0.120), magnesium(r =0.0006).

The porosity of the soil is negatively non significantly correlated with sulphur (r = -0.054), positively significantly correlated with phosphorus(r = 0.504) whereas positively non significantly correlated with water holding capacity(r = 0.255), OC (r = 0.028), pH(r = 0.180), EC(r = 0.319), nitrogen(r =0.233), potassium(r =0.134), calcium(r =0.252), magnesium(r =0.174).

The water holding capacity of the soil is negatively non significantly correlated with pH(r = -0.146), nitrogen(r = -0.155), sulphur(r =-0.232) and positively non significantly correlated with EC(r = 0.228), OC(r = 0.177), potassium(r = 0.238), magnesium(r = 0.027) and calcium(r = 0.246) whereas positively significantly correlated with phosphorus (r = 0.399)

The pH of the soil is negatively non significantly correlated with nitrogen(r = -0.105), organic carbon(r = -0.055) positively significantly correlated with calcium(r = 0.512) and positively non significantly correlated with EC(r = 0.375), phosphorus(r = 0.041), potassium(r = 0.291), magnesium(r = 0.226) and sulphur(r = 0.032). The EC of the soil positively non significantly correlated with nitrogen(r = 0.243), OC(r = 0.258), phosphorus(r = 0.176) and positively significantly correlated with potassium(r = 0.647), sulphur(r = 0.442), magnesium(r = 0.468), calcium(r =0.491).

The organic carbon content of soil is positively non significantly correlated with magnesium(r= 0.119) nitrogen(r = 0.077), phosphorus(r = 0.132), potassium(r = 0.189) and calcium(r = 0.227) whereas negatively non significantly correlated with sulphur (r =-0.172).

The micronutrients of the soil i.e. nitrogen status of the soil is negatively non significantly correlated with potassium(r = -0.042), phosphorus(r = -0.002) and magnesium(r = -0.262) and positively non significantly correlated with calcium(r = 0.005) and sulphur(r = 0.255). The phosphorus status of the soil is positively non significantly correlated with calcium(r = 0.352), magnesium(r =0.070). Positively significantly correlated with potassium(r = 0.416). Negatively significantly correlated with sulphur(r = -0.402). The potassium status of the soil is positively non significantly correlated with sulphur(r = 0.156). Positively significantly correlated with magnesium(r = 0.478) and calcium (r = 0.583).

The secondary macronutrients i.e. calcium content of the soil is negatively non significantly correlated with sulphur(r = -0.141) and positively non significantly correlated with magnesium(r = 0.106). The magnesium status of the soil is positively non significantly correlated with sulphur(r = 0.188).

Table: 4.1.5. Correlation between soil physico-chemical properties of Madanapalle town of Chittor district in Andhra Pradesh.

parameters	BD	PD	Porosity	WHC	pH	EC	OC	N	P	K	Ca	Mg	S
BD	1												
PD	0.128	1											
Porosity	-0.664*	0.652*	1										
WHC	-0.282	0.0366	0.255	1									
pH	-0.091	0.147	0.180	-0.146	1								
EC	-0.403*	0.007	0.319	0.228	0.375	1							
OC	-0.106	-0.035	0.028	0.177	-0.055	0.258	1						
N	-0.114	0.224	0.233	-0.155	-0.105	0.243	0.077	1					
P	-0.539*	0.139	0.504	0.399	0.041	0.176	0.132	-0.002	1				
K	-0.218	-0.043	0.134	0.238	0.291	0.647*	0.189	-0.042	0.416*	1			
Ca	-0.204	0.120	0.252	0.246	0.512*	0.491*	0.227	0.005	0.352	0.583*	1		
Mg	-0.210	0.0006	0.174	0.027	0.226	0.468*	0.119	-0.262	0.070	0.478*	0.106	1	
S	-0.096	-0.197	-0.054	-0.232	0.032	0.442*	-0.172	0.255	-0.402*	0.156	-0.141	0.188	1

Note: '' represents significant at 0.05 level*

4.2. Analysis of water parameters

4.2.1. The physical and chemical parameters of ground water

The data on the physical and chemical parameters of ground is given in Table-4.2 and in Appendix 7.

Values of pH in water samples ranged from 6.21 to 8.34 with 6.99 as a mean value. The SD and CV of the samples were 0.53 and 7.61% respectively. Sample no:22, site:8 Barlapalli village reported highest pH (pH 8.34) whereas sample no:1, site:1 Chipilli and sample no:23, site:8 Barlapalli village reported lowest (pH 6.21). Soil samples are neutral to alkaline in nature where only 2 samples (8.33 %) out of the total samples collected has shown moderately suitable for irrigation and 91.67 % of the samples are suitable for irrigation purpose. Higher pH is server to crops (Ayers and Westcot, 1985) and reduces the effectiveness of some nitrogenous fertilizers through irrigation (Bryan *et al.* 2007). Similar results were reported with Gummadi *et al.* (2015) in studying with irrigation water of coastal Andhra Pradesh.

Table: 4.2.1a. Classification of water samples under different pH range

Classes	Limit	No. of samples	% of samples
Suitable	6.5-8.4	22	91.6%
Moderately suitable	0-6.4	2	8.33%
Not suitable	>8.5	0	0%

(Source: WHO)

Values of electrical conductivity in water samples ranged between 0.50 to 4.32 dSm^{-1} with 1.62 dSm^{-1}as a mean value. The standard deviation of water samples is 1.17 and coefficient of variation is 72.4%. Sample no: 18, site: 6 Ramacharlapallli village reported highest (4.328 dSm^{-1}) EC while sample no: 11, site: 4 Kandlamadugu village reported lowest (0.504 dSm^{-1}) EC. 16.6% of the samples are in suitability range for irrigation, 62.5% of the samples are in moderate suitability range for irrigation and 20.8% of the samples are not suitable for irrigation. This clearly states that water quality in terms of EC in Madanapalle town are in moderate range for its use in irrigation. Higher EC concentration means high amount of total dissolved salts which causes water to lose its portability and reduces the solubility of oxygen in water (Kumar *et al.* 2011).

Table: 4.2.1b. Classification of water samples under different EC range

Classes	Limit (dS/m)	No. of samples	% of samples
Suitable	0.7	4	16.6%
Moderately suitable	0.7 – 3.0	15	62.5%
Not suitable	>3.0	5	20.8%

(Source: FAO.org)

4.2.2. Anions status in ground water

The data on anion status of ground water is given in the Table-4.2 and in Appendix 9.

The carbonate concentration in ground water samples of Madanapalle block are nil. The values of bicarbonate in water samples ranged between 1.4 to 4

Meq L^{-1} with 2.58 Meq L^{-1} as the average value. The SD and CV of the samples were 0.81 and 31.3% respectively. Sample no:18, site: 6 Ramacharlapallli and sample no:14, site 5 Basinikonda reported highest (4 Meq L^{-1}) concentration of bicarbonate while sample no:10 and 11 of site:4 Kandlamadugu and sample no:21, site:7 Ganganavaripalli village reported lowest (1.4 Meq L^{-1}) concentration of bicarbonate. Generally carbonate concentration of more than 0.1 Meq L^{-} and bicarbonate concentration of more than 10 Meq L^{-1} are not recommended for irrigation purposes according to Ayers and Westcot, 1994. It was found that 100% of the samples are in moderate range for irrigation. Similar findings were found with Acharya *et al.* (2010) in Gujarat.

Table: 4.2.2a. Classification of water samples under different alkalinity range

Classes	Limit	No. of samples	% of samples
Suitable	1.25	0	0%
Moderately suitable	1.25 – 8.50	24	100%
Not suitable	>8.50	0	0%

(Source: FAO.org)

The values of chloride concentration in water samples ranged from 2.4 to 30 Meq L^{-1} with 10.6 Meq L^{-1} as a mean value. The SD and CV of the samples were 9.17 and 91.1% respectively. Sample no: 7, site 3 Nayabandapalli reported lowest (2.4 Meq L^{-1}) concentration of chloride while sample no: 17, site: 6 Ramacharlapallli village reported highest (30 Meq L^{-1}) concentration of chloride. It was found that 37.5% of the samples are in suitable range, 33.33% of the samples are in moderately suitable range and 29.16% of samples are in not suitable for irrigation which conclude that the water samples from Madanapalle block are in moderate range for irrigation purpose.

Table: 4.2.2b. Classification of water samples under different chloride range

Classes	Limit	No. of samples	% of samples
Suitable	<4	9	37.5%
Moderately suitable	4 – 10	8	33.33%
Not suitable	>10	7	29.16%

(Source: FAO.org)

4.2.3. Cations status in ground water

The data on anion status of ground water such as calcium+magnesium, potassium and sodium are given in the Table-4.2 and in Appendix 8.

The values of calcium plus magnesium in water samples ranged from 4.8 to 58.4 Meq L^{-1} with 21.7 Meq L^{-1} as a mean value. The SD and CV of the samples were 14.4 and 66.1% respectively. Sample no: 6, site: 2 Ponnugutupalem reported lowest

(4.8 Meq L^{-1}) concentration of calcium plus magnesium while sample no: 13, site: 5 Basinikonda village reported highest (58.4 Meq L^{-1}) concentration of calcium plus magnesium. As per the guideline of ICMR, 1975 6.0 Meq L^{-1} is the highest desirable limit for total hardness. According to guideline it was found that only one sample i.e. sample no: 6 is less than 6.0 Meq L^{-1} and is in suitable range which states that water samples from Madanapalle block are not suitable for irrigation purpose in terms of hardness.

Values of potassium in water samples ranged from 0 to 0.512 Meq L^{-1} with 0.03 Meq L^{-1} as a mean value. The SD and CV of the samples were 0.01 and 51.6% respectively. Sample no: 1, 10 and 11 i.e. Chipilli and Kandlamadugu villages reported lowest (0 Meq L^{-1}) concentration of potassium while Chipilli, Ponnugutupalem, Nayabandapalli, Basinikonda, Ramacharlapallli, Ganganavaripalli and Barlapalli villages reported highest (0.512 Meq L^{-1}) concentration of potassium. As per the guideline of FAO, 1994 0-2 Meq L^{-1} is the range for potassium concentration in water samples. According to this data the potassium content in ground water samples of Madanapalle town are in moderately suitable range for irrigation purpose.

Values of sodium in water samples ranged from 1.39 to 4.78 Meq L^{-1} with 2.53 Meq L^{-1} as a mean value. The SD and CV of the samples were 1.13 and 44.5% respectively. Sample no: 11, site: 4 Kandlamadugu village reported lowest (1.391 Meq L^{-1}) concentration of sodium while sample no: 17 and 18 site: 6 Ramacharlapallli village reported highest (4.782 Meq L^{-1}) concentration of sodium. It was found that 70.83 % of the samples are in suitable range for irrigation and 29.16 % of the samples are in moderately suitable range for irrigation.

Table: 4.2.3a. Classification of water samples under different sodium range

Classes	Limit	No. of samples	% of samples
Suitable	<3	17	70.83%
Moderately suitable	3 – 9	7	29.16%
Not suitable	>9	0	0%

(Source: FAO.org)

4.2.4. Irrigation water quality index

The data on irrigation water quality index such as SAR, RSC, PI, SSP and KR are given in the Table-4.2 and in Appendix 10.

Irrigation water quality of index of samples ranges from 81.2-519.6 with a mean value of 225.5. The SD and CV of the samples were 140.2 and 62.1% respectively. The highest range is found in soil sample no: 18 Ramacharlapallli village and lowest range was found in soil sample no: 14 Basinikonda village. Out of total number of soil samples 62.5% of the samples are in poor range for irrigation, 4.16% only one sample is under very poor range for irrigation, 25% of the samples are unsuitable for irrigation and 8.33% of samples are in good range for its use for irrigation purpose. Due to presence of high calcium and magnesium i.e. total harness caused high water quality index in water.

Table: 4.2.4a. Classification of water samples under different IWQI range

WQI	Water quality status	No. of samples	% of samples
<50	Excellent	0	0%
50-100	Good	2	8.33
100-200	Poor	15	62.5%
200-300	Very poor	1	4.16%
>300	Unsuitable	6	25%

(Source: Anbazaghan, 2014)

Sodium adsorption ratio

The range of sodium adsorption ratio in water samples varies from 0.42 to 1.85 Meq L^{-1} with 0.85 Meq L^{-1} as a mean value. The SD and CV of the samples were 0.38 and 45.6% respectively. Sample no: 17, site: 6 Ramacharlapallli village reported highest (1.85 Meq L^{-1}) value of SAR while sample no: 23, site: 8 Barlapalli village reported lowest (0.42 Meq L^{-1}) value of SAR. It was found that 100 % of the samples are in permissible limit and belongs to class S1 which indicate no sodium hazard and ground water samples are in suitable range for irrigation purpose. Similar findings were found with Nag and Das (2015) in West Bengal.

Table: 4.2.4b. Classification of water samples under different sodium adsorption ratio range

Classes	Limit	No. of samples	% of samples
S_1(very low sodium hazard)	0-3	24	100%
S_2(low sodium hazard)	3-6	0	0%
S_3(medium sodium hazard)	6-12	0	0%
S_4 (high sodium hazard)	12-20	0	0%
S_5 (very high sodium hazard)	20-40	0	0%

(Source: FAO.org)

Soluble sodium percentage

The range of soluble sodium percentage in water samples varies from 5.67 to 29.5 with 13.1 as a mean value. The SD and CV of the samples were 7.74 and 58.7% respectively. Sample no: 21, site: 7 Ganganavaripalli village reported highest (29.57 %) value of SSP while sample no: 23, site: 8 Barlapalli village reported lowest (5.67 %) value of SSP. It was found that 100% of samples are in suitable range and fit for irrigation purposes.

Table: 4.2.4c. Classification of water samples under different soluble sodium percentage ratio range

Classes	Limit	No. of samples	% of samples
Good	<50	24	100%
Not suitable	>50	0	0%

(USDA, 1954)

Kelly ratio

The range of Kelly's ratio in water sample varies from 0.06 to 0.42 with 0.16 as a mean value. The SD and CV of the samples were 0.11 and 70.8% respectively. Sample no: 21, site: 7 Ganganavaripalli village reported highest (0.42) value of KR while sample no: 23, site: 8 Barlapalli village reported lowest (0.06) value of KR. It was found that 100% of the samples are in permissible range and fit for irrigation purposes.

Table: 4.2.4d. Classification of water samples under different range of Kelly's ratio range

Classes	Limit	No. of samples	% of samples
Good	<1.0	24	100%
Not suitable	>1.0	0	0%

(Source: Karanth, 1987)

Permeability index

The range of permeability index (PI) in water samples ranged from 9.65 to 50.9 with 21.9 as a mean value. The SD and CV of the samples were 12.08 and 54.9% respectively. Sample no: 6, site: 2 Ponnugutupalem village reported highest (50.91) value of PI while sample no: 15, site: 5 Basinikonda village reported lowest (9.65) value of PI. It was found that 33.33% of the samples are in moderately permissible limit and 66.66 % of the samples are not in permissible range and regarded as unfit for irrigation.

Table: 4.2.4e. Classification of water samples under different permeability index range

Classes	Limit	No. of samples	% of samples
Excellent	>75	0	0%
Good	25-75	8	33.33%
Not suitable	<25	16	66.66%

(Source: Doneen's, 1964)

Residual Sodium Bicarbonate

RSC of water samples ranged from -55.0 - -2.60 with a mean value of -19.2. The SD and CV of the samples were 14.0 and -73.3% respectively. In the collected water samples all the samples showed negative value of RSC indicating low risk of sodium accumulation due to offsetting levels of calcium and magnesium. Similar results were reported with Kumar *et al.* (2015) while studying the water quality of Lahor block. It was found that 100% of the samples are in permissible range and fall in low hazard category.

Table: 4.2.4f. Classification of water samples under different RSC range

Classes	Limit	No. of samples	% of samples
Low hazard	<1.25	24	100%
Medium hazard	1.25-2.25	0	0%
High hazard	>2.25	0	0%

(Source: Doneen's, 1964)

Table: 4.2. Statistical data on water parameters

Parameters	Mean	Range	S.D	C.V%
pH	6.98	6.21-8.34	0.53	7.61
EC	1.62	0.50-4.32	1.17	72.4
$Ca^{2+} + Mg^{2+}$	21.7	4.8-58.4	14.4	66.1
Ca^{2+}	13.1	3.8-39.8	8.15	61.8
K^{+}	0.03	0-0.512	0.01	51.6
Na^{+}	2.53	1.39-4.78	1.13	44.5
Cl^{-}	10.0	2.4-30	9.17	91.1
HCO_3	2.58	1.4-40	0.81	31.3
SAR	0.85	0.42-1.85	0.38	45.6
PI	21.9	9.65-50.9	12.0	54.9
KR	0.16	0.06-0.42	0.11	70.8
SSP	13.1	5.67-29.5	7.74	58.7
IWQI	225.5	81.2-519.6	140.2	62.1
RSC	-19.2	-55-2.6	14.08	-73.3

4.2.5. Correlation matrix between physico-chemical properties of water of different villages in Madanapalle town of Chittor district, Andhra Pradesh.

The data on correlation matrix between physico-chemical properties of water of different villages in Madanapalle town of Chittor district, Andhra Pradesh were presented in Table: 4.2.5a.

The pH of the water is positively non-significantly correlated with potassium (r = 0.149), bicarbonate(r = 0.194), sodium adsorption ratio (r = 0.023), Kelly's ratio (r = 0.010), permeability index (r = 0.080), soluble sodium percentage (r = 0.049). And negatively non significantly correlated with EC (r = -0.097), sodium (r = -0.022), calcium (r=-0.058), magnesium (r= -0.114), calcium+magnesium (r = -0.123), chloride (r = -0.121). The EC of the water is negatively non significantly correlated with permeability index (r = -018) and positively non significantly correlated with sodium (r = 0.981), potassium (0.204), magnesium (r = 0.318), bicarbonate (r = 0.776), chloride (r = 0.965), kelly's ratio(r = 0.150), soluble sodium percentage (r = 0.111), and positively significantly correlated with calcium+magnesium (r = 0.517). The sodium content of water is positively significantly correlated with calcium+magnesium (r = 0.487), sodium adsorption ratio (r = 0.647) and positively non significantly correlated with potassium (r = 0.249), calcium (r = 0.383), magnesium (r = 0.343), bicarbonate (r = 0.813), chloride (r = 0.931), permeability index (r = 0.028), kelly's ratio (r = 0.204), soluble sodium percentage (r = 0.165).The potassium content of water is positively non significantly correlated with calcium(r = 0.121), calcium+magnesium (r = 0.146), magnesium (r = 0.098), bicarbonate (r = 0.388), chloride (r = 0.118), sodium adsorption ratio (r = 0.150), permeability index (r = 0.005), Kelly's ratio (r = 0.047), soluble sodium percentage(r = 0.009). The calcium content of the water is positively significantly correlated with calcium+magnesium (r = 0.616), chloride (r = 0.450) while positively non significantly correlated with magnesium (r = 0.616), bicarbonate (r = 0.309) and negatively significantly correlated with permeability index (r= -0.568), Kelly's ratio (r = -0.478), soluble sodium percentage (r = -0.483) while negatively non significantly correlated with sodium adsorption ratio (r = -0.182), The magnesium content of the water is positively non significantly correlated with calcium+magnesium (r = 0.825), bicarbonate (r = 0.311), chloride (r = 0.315) while negatively significantly correlated with permeability index (r = -0.525), Kelly's ratio (r = -0.452), soluble sodium percentage (r = -0.479) and negatively non significantly correlated with sodium adsorption ratio (r= -0.254). The calcium+magnesium content of the water is negatively non significantly correlated with sodium adsorption ratio (r = -0.304), permeability index (r = -0.736), while negatively significantly correlated with kelly's ratio (r = -0.627), soluble sodium percentage (r = -0.651), and positively significantly correlated with bicarbonate (r = 0.420), chloride (r = 0.503).The bicarbonate content of the water is positively significantly correlated with sodium adsorption ratio (r = 0.429) while positively non significantly correlated with chloride (r = 0.724), kelly's ratio (r = 0.001), and negatively non significantly correlated with permeability index (r = -0.088), soluble sodium percentage (r = -0.038). The chloride content of the water is positively significantly correlated with sodium adsorption ratio (r = 0.559) while positively non significantly correlated with kelly's ratio (r = 0.139), soluble sodium percentage (r = 0.102) and negatively non significantly correlated with permeability index (r = -0.016).The sodium adsorption ratio of water is positively non significantly correlated with kelly's ratio (r = 0.862), permeability index (r = 0.748), soluble sodium percentage (r = 0.827). The permeability index of water is positively non significantly correlated with Kelly's ratio (r= 0.965), soluble sodium percentage (r = 0.972). The Kelly's ratio of the water is positively non significantly correlated with soluble sodium percentage (r = 0.984).

Table: 4.2.5a. Correlation between water physico-chemical properties of Madanapalle town of Chittor district in Andhra Pradesh.

Parameters	pH	EC	Na^+	K^+	Ca^{2+}	Mg^{2+}	Ca+Mg	HCO_3	Cl^-	SAR	PI	KR	SSP	RSC
pH	1													
EC	-0.097	1												
Na^+	-0.022	0.981	1											
K^+	0.149	0.204	0.249	1										
Ca^{2+}	-0.058	0.470*	0.382	0.121	1									
Mg^{2+}	-0.114	0.318	0.343	0.098	0.064	1								
Ca+Mg	-0.123	0.517*	0.487*	0.146	0.616*	0.825	1							
HCO_3	0.194	0.776	0.813	0.388	0.309	0.311	0.420*	1						
Cl^-	-0.121	0.965	0.931	0.118	0.450*	0.315	0.503*	0.724	1					
SAR	0.023	0.593*	0.647*	0.150	-0.182	-0.254	-0.304	0.429*	0.559*	1				
PI	0.080	-0.018	0.028	0.005	-0.568*	-0.525*	-0.736	-0.088	-0.016	0.748	1			
KR	0.010	0.150	0.204	0.047	-0.478*	-0.452*	-0.627*	0.001	0.139	0.862	0.965	1		
SSP	0.049	0.111	0.165	0.009	-0.483*	-0.479*	-0.651*	-0.038	0.102	0.827	0.972	0.984	1	
RSC	0.137	-0.484	-0.451	-0.127	-0.612	-0.826	-0.998	-0.373	-0.473	0.335	0.748	0.642	0.664	1

Note: '' represents significant at 0.05 level.*

5

CHAPTER

Summary and Conclusion

A research was conducted on the assessment of soil and water quality status entitled "Study of Physico-Chemical parameters of soil and water in Madanapalle block of Chittor district, Andhra Pradesh". The study was carried out during July-September 2019 where eight different villages were selected in Madanapalle block and from each village 3 surface (0-15cm) soil and water samples were taken and analysed for Physical and chemical parameters using standard laboratory protocols. The correlation matrix between soil and water parameters were also briefed. The salient findings of the present investigation are summarized under:

5.1 Soil samples analysis of Madanapalle block

- The range of Bulk density and Particle density in the soil samples varied from 1.16- 1.43 Mg/m^3 and 2.08 to 2.67 Mg/m^3 with 1.30 and 2.37 Mg/m^3 as the mean values.
- The range of Porosity percentage in the soil samples varied from 38.3 to 51.8 with 44.9% as the mean value and Water holding capacity in the soil samples varied from 32.5 to 60.6 with 47.1% as the mean value.
- The pH of the soil samples are in alkaline to neutral in condition and values ranged from 5.87 to 8.50 with 7.54 as a mean value. The present findings concluded that alkaline nature of soil samples are in percentile of 62.5%, neutral nature of soil samples are in percentile of 29.16% and acidic nature of soil samples are in 8.32%.
- The range of Electrical Conductivity in the soil samples varied from 0.072 to 1.264 dSm^{-1} with 0.45 dSm^{-1} as a mean value. It was observed that 83.3% of the samples are in permissible range suitable for all type of crops and 16.6% of the samples are slightly higher than the permissible range suitable for most of the crops.
- The range of organic carbon in the soil samples varied from 0.24 to 0.95 with 0.59 as a mean value. The present findings concluded that high range of organic carbon was found in 62.5% of the soil samples, low range of organic carbon was found in 8.33% of the soil samples and medium range of organic carbon was found in 29.1% of the soil samples.
- The available Nitrogen content in soil samples are in medium to low range. The range of available nitrogen in the soil samples varied from 72.31 to 765.9 Kg ha^{-1} with 248.6 Kg ha^{-1} as a mean value. The present findings

concluded that low range of nitrogen content was found in 70.83% of the soil samples, medium range of nitrogen content was found in 16.6% of the soil samples and high range of nitrogen content was found in 12.5% of the soil samples.

- The available phosphorus content in soil samples are in medium to low range. The range of available phosphorus in the soil samples varied from 1.25 to 31.2 Kg ha^{-1} with 13.7 Kg ha^{-1} as a mean value. The present findings concluded that low range of phosphorus content was found in 54.1% of the soil samples, medium range of phosphorus content was found in 29.1% of the soil samples and high range of phosphorus content was found in 16.6% of the soil samples.
- The available potassium content in soil samples are in medium to high range. The range of available potassium in the soil samples varied from 89.6 to 1448.8 Kg ha^{-1} with 481.7 Kg ha^{-1} as a mean value. The present findings concluded that low range of potassium content was found in 8.3% of the soil samples, medium range of potassium content was found in 25% of the soil samples and high range of potassium content was found in 66.66% of the soil samples.
- The available sulphur content in soil samples are in low range. The range of available sulphur in the soil samples varied from 0.14 to 0.45 mg kg^{-1} with 0.25 mg kg^{-1} as a mean value. The present findings concluded that in Madanapalle town there is low concentration of sulphur where low range of sulphur content was found in 100% of the soil samples.
- The exchangeable calcium content in soil samples are in high range. The range of exchangeable calcium in the soil samples varied from 0.40 to 9.08 Meq/100g with 2.96 Meq/100g as a mean value. The present findings concluded that low range of calcium content was found in 37.5% of the soil samples and high range of calcium content was found in 62.5% of the soil samples.
- The exchangeable magnesium content in soil samples are in low range. The range of exchangeable magnesium in the soil samples varied from 0.04 to 3.70 Meq/100g with 1.05 Meq/100g as a mean value. The present findings concluded that low range of magnesium content was found in 58.3% of the soil samples and high range of magnesium content was found in 41.6% of the soil samples.
- The nutrient index range of soil samples were high for organic carbon, low for nitrogen, low for phosphorus, high for potassium, and low for sulphur.
- 5.2 Water samples analysis of Madanapalle block
- The pH of the groundwater samples are in neutral to slightly alkaline in condition and values ranged from 6.21 to 8.34 with 6.99 as a mean value. The present findings concluded that moderately suitable range of pH was found in only 2 samples (8.33 %) and suitable range of pH was found in 91.67 % of the samples.

- The range of EC in the water samples varied from 0.50 to 4.32 dSm^{-1} with 1.62 dSm^{-1} as the mean value. The present findings concluded that suitable range of EC was found in 16.6% of the samples, moderately suitable range of EC was found in 62.5% of the samples and not suitable range of EC was found in 20.8% of the samples.
- The range of chloride concentration in the water samples varied from 2.4 to 30 Meq L^{-1} with 10.6 Meq L^{-1} as the mean value. The present findings concluded that suitable range of chloride was found in 37.5% of the samples, moderately suitable range of chloride was found in 33.33% of the samples and not suitable range of chloride was found in 29.16% of the samples.
- The range of bicarbonate concentrations in the water samples varied from 1.4 to 4 Meq L^{-1} with 2.58 Meq L^{-1} as the mean value. The range on potassium concentration in the water samples varied from 0 to 0.512 Meq L^{-1} with 0.03 Meq L^{-1} as the mean value.
- The range of sodium concentration in the water samples varied from 1.39 to 4.78 Meq L^{-1} with 2.53 Meq L^{-1} as the mean value. The present findings concluded that suitable range of sodium was found 70.8% of the samples, moderately suitable range of sodium was found in 29.16% of the samples. The range of calcium plus magnesium in the water samples varied from 4.8 to 58.4 Meq L^{-1} with 21.7 Meq L^{-1} as the mean value.
- The range of irrigation water quality index (IWQI) in the water samples varied from 81.2-519.6 with 225.5 as the mean value and has shown that 62.5% of the samples are in poor range for irrigation, 4.16% only one sample is under very poor range for irrigation, 25% of the samples are unsuitable for irrigation and 8.33% of samples are in good range for its use for irrigation purpose.
- In the water samples the SAR values ranged from 0.42 to 1.85 Meq L^{-1} with 0.85 Meq L^{-1} as the mean value. It was found that 100 % of the samples are in permissible limit and belongs to class S1 which indicate no sodium hazard and ground water samples are in suitable range for irrigation purpose.
- The range of residual sodium carbonate in the water samples is from -55.0 - -2.60 with -19.2 as the mean value. SSP, KR and PI concentration in water samples ranges from 5.67 to 29.5, 0.06 to 0.42 and 9.65 to 50.9 with a mean value of 13.1, 0.16 and 21.9.

Conclusion

The soil analysis results were interpreted using the literature which help the farmers to analyse and add the deficient nutrients. According to the soil test results of Madanapalle block clearly states that the soil is alkaline to neutral in condition. 50% of the soil samples showed medium-high organic carbon content this is due prevailing low temperature and less decomposition of organic matter in the soil. The potassium content of the soil was medium to high in condition, so the soils are efficient in potassium content in Madanapalle block. By considering nutrient index the soils of Madanapalle block are low in nitrogen, phosphorus and sulphur whereas high in potassium and organic carbon.

Due to usage of high pesticides and fertilizers by the farmers and also due to inappropriate cultivation practices there was accumulation of soluble salts which resulted moderate range of EC concentration in the samples. The pH was found to be neutral to slightly alkaline in nature and all the water samples have shown low sodium hazard and alkalinity of all the water samples have shown moderately suitable for irrigation. The permeability index of water showed 66.66% of samples are not suitable for irrigation purpose and 33.33% of samples are moderately suitable. The calcium and magnesium concentration of the samples were quite high. Due to high concentration calcium and magnesium and moderate range of soluble salts in water samples indicated poor quality irrigation water index where 62.5% of the samples were in poor range. This results shows that still improvement has to be done to improve the quality of irrigation water mainly by improving the cultivation practices. With the knowledge and experience gained through study may be developed in future to help the farmers regarding the quality of produce, high yields through soil and water conservation and better environmental protection.

Bibilography

Acharya. G. D., Solanki. M. R. and Hathi M. V. (2010). Studies on Physico-Chemical Parameters of Irrigation Water, Prantij, Gujarat (India). *International Journal Chemical Sciences,* **8**(4): 2377-2385.

Ahad.T., Kanth. T.A. and Nabi. S. (2015). Soil Bulk Density as Related to Texture, Organic Matter Content and Porosity in Kandi Soils of District Kupwara (Kashmir Valley), India. *International Journal of Scientific Research,* **4**(1): 198-200.

Ali Khan. S.M., Mahmood. S., Chaudhary. M.N. and Sadiq. M. (2009). Irrigation quality of groundwater of twenty villages in Lahore district. *Journal of Soil and Environment,* **28**(1): 17-23.

Allen., Barker. V. and Pilbeam. (2007). Hand book of plant nutrition. *CRC press,* 2: 1-773.

Alper. G. and Hakan. A. (2016). Assessment of Water Quality in Drainage Canals of Carsamba Plain, Turkey, through Water Quality Indexes and Graphical Methods. *Global NEST Journal,* **18**(1): 67-78.

Amanial. (2016). Physico-Chemical Analysis of Drinking Water Quality of Arbaminch Town. *Journal of Environmental & Analytical Toxicology,* **6**(2): 1-5.

Anbazhagan. S., Das. M.K. and Balamurugan. G. (2014). Hydro geo-chemical studies in hard rock aquifer system, Hosur, India. *Indian Journal of Geo-Chemistry,* **22**(2): 519-535.

APHA. (1992). Standard methods for the examination of water and waste waters. *American Public Health and Association,* 18th Edition, Washington.

Ayers. R.S. and Westcot. D.W. (1985). Water quality for agriculture. *Rome FAO,* 29: 37-70.

Babiker.S., Mohamed. A. and Hiyama T. (2007). Assessing the groundwater quality using GIS. *Water resource management,* **21**(4): 699-715.

Basavaraja. P.K., Dey. P., Saqeebulla. M.H. and Yogendra. N.D (2017). Geo-reference based soil fertility status in Hassan district of Karnataka, India for development of nutrient plan. *Indian Journal of Soil Conservation,* **45**(2): 141-147.

Bhat. A.M., Grewal. M.S., Prakash. R. and Yadav. R. (2016). Assessment of groundwater quality for irrigation purposes using chemical indices. *Indian Journal of Ecology,* **43**(2): 574-579.

BIS (1991). Indian drinking water standards. First revision, edition 2.1.

Black. C. A. (1965). Methods of Soil Analysis. Part I and II. American Society of Agronomy, Inc., Madison, Wisconsin, U. S. A., 1-2: 1572.

Bryan. G., Hopkins., Donald. A., Horneck. R.G., Stevens. J., Ellsworth. W., Dan. M. and Sullivan (2007) Managing Irrigation Water Quality, for crop production in the Pacific Northwest, *A Pacific Northwest Extension publication PNW*, 597-E.

Chakraborty. S. and Nandini. V. (2015). Drinking water Quality assessment for pH and Total Dissolved Solids. *International Research Journal of Environmental Sciences*, **4**(8): 39-44.

Chandak. N., Shabana. M.B.P., Meena. D. and Kamlesh. S. (2017). Analysis of Soil Samples for its Physico-Chemical Parameters from Kadi City. *Newest International Referred Journals*, 1: 2349-3372.

Chandu. S.N., Sibbarao. N.V. and Prakash. S.R. (1995). Suitability of ground water for domestic and irrigational purposes in some parts of Jhansi District, U.P. *Bhujal News*, 10(1): 12-17.

Chaudhari. K.G. (2013). Studies of the physicochemical parameters of soil samples. *Advances in Applied Science Research*, **4**(6): 246-248.

Chaudhari. P.R., Ahire. D.V., Ahire. V.D., Chakravarty. M. and Maity. S. (2013). Study of Soil bulk density as related to soil texture, organic matter contents and available total nutrients of Coimbatore soil. *International Journal of Scientific and Research publications*, **3**(2): 1-8.

Chaudhary. V. and Kumar. S. (2018). Assessment of groundwater quality for drinking and irrigation purposes in arid areas of Rajasthan, India. *Applied Water Science*, 8: 218-235.

Chesnin. L. and Yien. C.H. (1950). Turbidimetric determination of available sulphur. *Proceeding of Soil Science. American*, 14: 149-151.

Chesworth. W. (2008). Encyclopaedia of Soil Science. *Springer*, 1: 1-928.

Dameshwar. M. and Navaz. (2018). Evaluation of soil fertility status of Matiya village of Kasdol block under Balodabazar district of Chhattisgarh. *International Journal of Chemical Studies*, **6**(2): 38-41

Das. M.V. and Poongothai. S. (2014). Spatial Analysis of Soil and Water Quality in Tsunami Affected Areas of Nagapattinam District, Tamilnadu, India. *Journal of Engineering Research and Applications*, **4**(9): 167-173.

Deb P., Debnath. P., and Pattanaaik S.K. (2013). Physico-chemical properties and water holding capacity of cultivated Soils along altitudinal gradient in South Sikkim, India. *Indian Journal of Agricultural Research*, **48**(2): 120-126.

Doneen. L.D. (1964). Water quality for agriculture. Department of irrigation, *University of California*, p: 48.

Eaton. F.M. (1950) Significance of carbonate in irrigation water. *Soil Science*, 69: 123–133.

FAO (1994). Water quality for Agriculture FAO irrigation drainage paper, No: 29, rev.1.Food and Agriculture organization of the United Nations, Rome.

Gothwal. R. and Kumar. G. G. (2018). Limnological study of Lentic Fresh Water Ecosystem during Summer Season: Nakki Lake-Mount Abu, Rajasthan, India. *An International Scientific Journal,* 114: 44-54.

Gummadi. S., Swarnalatha. G., Brahmaji Rao. P., Venkataratnamma. V. and Vijaya Kumar. G. (2015). Study of Irrigation Water Quality With Reference To Coastal Andhra Pradesh, India. *International Journal of modern Engineering research,* **5**(4): 55-60.

Indian Council of Medical Research (ICMR) (1975). Manual of Standards of Quality for Drinking Water Supplies. Special Report Series No. 44. New Delhi.

Jackson. M. L. (1973). Soil Chemical Analysis. *Prentice Hall of India Pvt. Ltd.,* New Delhi.

Jain S.A., Jagtap. M.S. and Patel. K.P. (2014). Physico-Chemical Characterization of farmland Soil used in some villages of Lunawada Taluka. Dist: Mahisagar (Gujarat) India. *International Journal of Scientific and Research Publications,* **4**(3): 1-5.

Kadam. P. M. and Balasaheb. U. K. (2016). Physicochemical Properties and its Relationship with Water Holding Capacity of Soil in Deulgaon Raja region *Maharashtra. Science and Technology,* **2**(6): 49-52.

Kant. S., Singh Y.V., Lokesh Jat. K., Meena. R. and Singh. S.N (2015). Assessment of ground water quality of lahar block, Bhind district in Madhya Pradesh. *International Journal of Advanced Geosciences,* **3**(2): 38-41.

Kanwar. J.S. (1976). Soil Fertility: Theory and Practice. S.S. Grewal, Under Secretary, for the ICAR, New Delhi.

Karanth. K. R. (1987). Ground Water Assessment Development and Management (New York: McGraw-Hill), pp.52-89.

Kelly. W. P. (1953). Use of Saline Irrigation Water. *Soil Science,* **95**(6): 385-391.

Khichi. Y. (2018). Narmada river water quality assessment using benthic macro-invertebrates at Barwani, Rajghat Madhya Pradesh. *Bioscience Biotechnology Research Communications,* **11**(1): 161-166.

Kumar. A.S., Shirur.M. and Sharma. V.P. (2017). Assessment of Soil Fertility Status of Mid Himalayan Region, Himachal Pradesh. *Indian Journal of Ecology,* **44**(2): 226-231.

Kumar. S.K. and Kumar. R. (2011). Analysis of water quality parameters of groundwater near Ambattur industrial area, Tamil Nadu, India. *Indian Journal of Science and Technology,* **4**(5): 660-662

Kumar.S.V., Amarender. B., Dhakate. R., Sankaran. S. and Kumar.R. K. (2016). Assessment of groundwater quality for drinking and irrigation use in shallow hard rock aquifer of Pudunagaram, Palakkad District Kerala. *Journal of Applied Water Sciences,* 6:149–167.

Marimuthu Krishnaveni. (2015). Nutrient Analysis of Soil Collected from Panuchakuli Village, Kanyakumari District, Kanyakumari. *Research Journal of Pharmacy and Technology,* **8**(7): 857-859.

Mayas. A., Esmail. S. A. and Sameer A. (2015). Groundwater Quality at Maytam Area in the City of Ibb, Yemen by Using Physico-chemical Analysis. *International Journal of Emerging Technology and Advanced Engineering,* **5**(3): 40-46.

Md. Rahman. M., Md. Islam. M., Md. Bodrud-Doza., Md. Muhib I., Shammi M., Zahid A., Akter. Y. and Kurasaki M. (2017). A study of groundwater irrigation water quality in south-central Bangladesh a geo-statistical model approach using GIS and multivariate statistics. *Springer-Verlag GmbH,* **37**(2): 193-214.

Mehta. A., Senapati. T. and Duggal. R. (2014). Study on Hydro geochemistry of Groundwater in Sanganer Tehsil, Jaipur District, Rajasthan. *International Journal of Geology, Earth & Environmental Sciences,* **4**: 183-193.

Meireles., Andrade., Chaves., Frischkorn. H. and Crisostomo. (2010). A new proposal of the classification of the Irrigation Water. *Revista Ciencia Agronomica,* **41**(3), 349–357

Mendonsa. L. and Vishnuprasad. V. (2019). Comparative Ecological Analysis of Five Freshwater Lakes in and Around Mumbai, India. *Journal of Ecology & Natural Resources,* **3**(5): 00178.

Mirza. A.T.M., Rahman. T., Rahman. S.H. and Majumder. R.K. (2012). Ground water quality for irrigation of deep aquifer in Southern-West zone of Bangladesh. *Journal of Science and Technology,* **34**(3): 345-352.

Muhr. G. R., Datta. N. P., Subramany. N. S., Dever. F., Lecy. V. K. and Donahue, R. R. (1963).Soil testing in India. *USDA Publication,* pp: 120.

Nag. S.K and Das.S. (2014). Quality Assessment of Groundwater with Special Emphasis on Irrigation and Domestic Suitability in Suri I & II Blocks, Birbhum District, West Bengal, India. *American Journal of Water Resources,* **2**(4): 81-98.

Olsen. S.R., Cole. C.V., Watanabe. F.S. and Dean. L.A. (1954). Estimation of Available Phosphorus in Soils by Extraction with Sodium Bicarbonate. *U. S. Department of Agriculture,* Circular No. 939.

Pandey. N.C., Lalit. M., Tewari., Joshi. G.C., and Upretin M.B. (2018). Physico-chemical characterization of Oak, Pine and Sal forest soil profiles of Betalghat Region of Kumaun Himalaya. *Eurasian Journal of Soil Science,* **7**(3): 261-272.

Pandey. V., Gautam. P. and Singh. A.P. (2018). Assessment of physical properties of soil under different land use systems in a Mollisol. *Journal of Pharmacognosy and Phytochemistry,* **7**(6): 2645-2648.

Parker. F.W., Nelson. W.L., Winters. E. and Miles. A.E. (1951). The board interpretation and application of soil test summaries. *Agronomy Journal,* **43**(3): 103-112.

Periyaswamy. K., Dahms. H.U., Jeon. H., Rajendran. A. and James. A.R. (2015). Irrigation water quality assessment from the Tamirapani River, Southern India. *Arab Journal of Geosciences,* **7**: 5209-5220.

Piper, C. S. (1966).Soil and Plant analysis. *Hans Publisher,* Bombay.

Ramamurthy. B, Bajaj. J.C. (1969). Nitrogen, Phosphorus and Potash status of Indian soils. *Fertilizer News,* 14: 25-28.

Richards. L. A. (1954). Diagnosis and Improvement of Saline and Alkali Soils. *U. S. Department of Agriculture Handbook,* Washington D. C., USA. 60: 160.

Sanda. A.R. and Dibal. J. (2014). Analysis of Irrigation Water Quality at Kadawa Irrigation Project for Improved Productivity. *International journal of environment,* **3**(3): 235-240 .

Schollenberger. C. J. and Simon. R. H. (1945). Determination of Exchange Capacity and Exchangeable Bases in Soil-Ammonium Acetate Method. *Soil Science,* **59**: 13-24.

Sharma. R., Madhuri. S., Rishi. L.R. and Herojeet. R. (2014). Evaluation of Surface Water Quality for irrigation purposes in Limestone Mining areas of District Solan, Himachal Pradesh, India. *International Journal of Innovative Science, Engineering & Technology,* **1**(8): 369-375.

Shivanna. A.M. and Gopalpur. N. (2014). Chemical Analysis of Soil Samples to Evaluate the Soil Fertility Status of Selected Command Areas of Three Tanks in Tiptur Taluk of Karnataka, India. *IOSR Journal of Applied Chemistry,* **7**: 01-05.

Shukla. M. and Arya. S. (2018). Determination of Chloride Ion (Cl-) Concentration in Ganga River Water by Mohr Method at Kanpur, India. *Green Chemistry & Technology Letters,* **4**(1): 06-08.

Singh. S.P., Singh. S., Ashok. K. and Rajesh. K. (2018). Soil Fertility Evaluation for Macronutrients Using Parkers Nutrient Index Approach in Some Soils of Varanasi District of Eastern Uttar Pradesh, India. *International Journal of Pure Applied Biosciences,* **6**(5): 542-548.

Singh. Y.V., Shashi Kant., Singh S.K., Sharma P.K., Jat., Kumar.M., Shahi. S.K., Jatav. H. S. and Yadav. R.N. (2017). Assessment of Physico-Chemical Characteristics of the Soil of Lahar Block in Bhind District of Madhya Pradesh (India). *International Journal of Current Microbiology and Applied Sciences,* **6**(2): 511-519.

Soni. M. (2016). Analysis of soil samples for its Physico-Chemical parameters from Abohar city. *The Pharma Innovation Journal,* **5**(11): 37-39.

Subbiah. B. V. and Asija. G. L. (1956).A rapid procedure for the determination of available nitrogen in soils. *Current Science,* 25: 259-260.

Todd. D.K. (1980). Groundwater Hydrology, Second Edition, Wiley, New York.

Twum. E. K.A. and Annang- SethNii (2015). Impact of Soil Compaction on Bulk Density and Root Biomass of Quercus petraea L. at Reclaimed Post-Lignite Mining Site in Lusatia, Germany. *Hindawi Publishing Corporation Applied and Environmental Soil Science Volume,* 1-5.

USDA (1954).U.S. Salinity Laboratory Staff, Diagnosis and Improvement of Saline and Alkali Soils. *Handbook 60.* Washington, D.C.

Vijayakumar, R., Arokiaraj.A. A. and Prasath, P.M.D. (2011).Micronutrients status and their Relation to Soil characteristics of South-East Coastal Soils of India. *International Journal of Research in chemistry and environment,* **1**(1): 8-12.

Viji. R. and Rajesh P.P. (2012). Assessment of water holding capacity of major soil series of Lalgudi, Trichy, India. *Journal of Environmental Research and Development,* **7**(1): 393-398.

Wagh. G. S., Chavhan. D. M. and Sayyed. M. R. G. (2013). Physicochemical Analysis of Soils from Eastern Part of Pune City. *Universal Journal of Environmental Research and Technology,* **3**(1): 93-99.

Walkley. A. and Black. T. A. (1934). An examination of the Degt. Jarett method for determination of soil organic matter and a proposed modification of chromic acid titration. *Soil Science,* **37**: 29-38.

World Health Organization (WHO) (2011). Guidelines for Drinking-Water Quality, WHO Press, Geneva, Switzerland, 4th edition.

Appendices

Appendix-1: Description of Soil sampling sites

Village name	Sample No.	Farmers name	Land type	Latitude	Longitude	Altitude
Site 1: Chipilli	S_1	Venkataraman	Cultivated tomato	13.5561S	78.48013E	709(±3m)
	S_2	Rambabu	Fallow land	13.55950N	78.48114W	717(±3m)
	S_3	Seena	Fallow land	13.56026S	78.48149W	721(±3m)
Site2: Ponnugutupalem	S_4	Ramireddy	Cultivated tomato	13.56047	78.48655E	710(±3m)
	S_5	Ravindranath	Fallow land	13.56194N	78.48207	720(±4m)
	S_6	Shiva reddy	Fallow land	13.56198S	78.48165	718(±3m)
Site:3 Nayabandapalli	S_7	Soma Shekar	Cultivated mango	13.6690S	78.4226E	718(±3m)
	S_8	Raja reddy	Grassed land	13.66880S	78.42162	721(±4m)
	S_9	Ravi chadra	Cultivated mango	13.67134N	78.42041	719(±3m)
Site:4 Kandlamadugu	S_{10}	S.A. Mastan	Fallow land	13.66824N	78.41873W	724(±3m)
	S_{11}	Pullaiah	Cultivated mango	13.66555S	78.41940W	725(±4m)

Village name	Sample No.	Farmers name	Land type	Latitude	Longitude	Altitude
	S_{12}	Triveni	Cultivated mango	13.66466N	78.41672E	741(±3m)
Site:5 Basinikonda	S_{13}	Susheel	Fallow land	13.55242N	78.52216	680(±3m)
	S_{14}	Naveen	Fallow land	13.55423N	78.51842W	676(±3m)
	S_{15}	Bharath	Fallow land	13.55285N	78.51864E	675(±3m)
Site:6 Ramacharlapallli	S_{16}	Siraj	Cultivated tomato	13.54999	78.53031W	676(±4m)
	S_{17}	Balu Swamy	Cultivated marigold	13.55068S	78.53369E	667(±4m)
	S_{18}	Ghouse	Fallow land	13.55101S	78.53814W	663(±3m)
Site:7 Ganganavaripalli	S_{19}	Venkatesh	Fallow land	13.54892	78.55058W	659(±3m)
	S_{20}	Vishnu	Fallow land	13.54892S	78.55048E	665(±3m)
	S_{21}	Chinnapa	Cultivated tomato	13.54969S	78.55230	671(±3m)
Site:8 Barlapalli	S_{22}	Shankar reddy	Cultivated gooseberry	13.54149	78.43281	744(±3m)
	S_{23}	Kallu Banda	Fallow land	13.54306	78.43305	748(±3m)
	S_{24}	Talla reddy	Fallow land	13.54706	78.43575	751(±3m)

Appendix-2: Status of Bulk density, Particle density, Porosity and WHC in soil

Sample No	Bulk density (g/ cm³)	Particle density(g/ cm³)	Porosity (%)	Water holding capacity (%)
S_1	1.24	2.08	40.2	45.9
S_2	1.36	2.41	43.4	41.6
S_3	1.33	2.38	44.2	50.9
S_4	1.36	2.30	41.0	41.0
S_5	1.39	2.26	38.3	36.8
S_6	1.36	2.36	42.2	32.5
S_7	1.43	2.33	38.6	47.5
S_8	1.32	2.39	44.8	35.4
S_9	1.26	2.34	46.2	46.5
S_{10}	1.28	2.28	43.8	51.7
S_{11}	1.26	2.20	42.9	47.5
S_{12}	1.29	2.41	46.5	53
S_{13}	1.27	2.22	42.7	46.7
S_{14}	1.30	2.67	51.2	54.4
S_{15}	1.25	2.35	46.7	55.7
S_{16}	1.43	2.43	41.1	60.6
S_{17}	1.29	2.41	46.3	41.6
S_{18}	1.25	2.52	50.2	36.8
S_{19}	1.27	2.38	46.5	53.2
S_{20}	1.16	2.40	51.8	48.3
S_{21}	1.33	2.54	47.4	47.7
S_{22}	1.35	2.55	46.9	39.9
S_{23}	1.19	2.43	50.8	57.9
S_{24}	1.23	2.27	45.9	59.5
MEAN	1.30	2.37	44.98	47.19
RANGE	1.16-1.43	2.08-2.67	38.3-51.8	32.5-60.6
SD±	0.06	0.12	3.77	7.81
CV (%)	5.27%	5.28%	8.39%	16.5%

Appendix-3: Status of available pH, EC and organic carbon in soil.

Sample No	pH	Electrical conductivity(dSm^{-1})	Organic carbon (%)
S_1	7.82	0.326	0.95
S_2	7.84	0.288	0.65
S_3	8.11	0.297	0.42
S_4	7.96	0.664	0.72
S_5	6.65	0.241	0.31
S_6	8.16	0.072	0.24
S_7	6.89	0.216	0.83
S_8	7.68	0.339	0.53
S_9	7.86	0.395	0.74
S_{10}	7.46	0.230	0.47
S_{11}	6.78	0.371	0.64
S_{12}	7.56	0.259	0.39
S_{13}	8.01	0.937	0.77
S_{14}	8.47	0.806	0.91
S_{15}	7.63	0.676	0.68
S_{16}	7.23	0.309	0.67
S_{17}	7.72	0.707	0.43
S_{18}	7.12	0.146	0.86
S_{19}	7.65	1.264	0.72
S_{20}	7.90	1.221	0.57
S_{21}	5.87	0.107	0.71
S_{22}	8.5	0.342	0.24
S_{23}	7.45	0.243	0.48
S_{24}	6.72	0.387	0.38
MEAN	7.54	0.451	0.59
RANGE	5.87-8.50	0.072-1.264	0.24-0.95
SD±	0.62	0.32	0.20
CV (%)	8.24%	72.7%	34.5%

Appendix-4: Status of available primary macronutrient in soil.

Sample No	Nitrogen (kg ha^{-1})	Phosphorus (kg ha^{-1})	Potassium (kg ha^{-1})
S_1	75.4	14.6	268.8
S_2	132.2	5.38	403.2
S_3	97.0	3.50	436.8
S_4	144.5	1.25	257.6
S_5	765.9	3.38	392
S_6	178.9	11.7	268.8
S_7	413.2	3.25	448
S_8	134.2	8.26	347.2
S_9	154.3	10.6	638.4
S_{10}	117.8	5.44	459.2
S_{11}	158.5	19.5	739.2
S_{12}	84.7	31.1	728
S_{13}	187.4	18.6	929.6
S_{14}	564.7	31.1	896
S_{15}	195.4	31.2	728
S_{16}	145.6	4.25	89.6
S_{17}	342.1	13.0	459.2
S_{18}	523.4	23.2	179.2
S_{19}	154.4	17.0	1448.8
S_{20}	732.3	7.07	369.6
S_{21}	125.7	7.51	89.6
S_{22}	72.3	9.07	414.4
S_{23}	342.1	20.0	257.6
S_{24}	126.7	29.1	313.6
MEAN	248.7	13.71	481.7
RANGE	72.3-765.9	1.25-31.2	89.6-1448.8
SD±	204.8	9.77	308.6
CV (%)	82.3%	71.2%	64.0%

Appendix-5: Status of available secondary macronutrients in soil.

Sample No	Calcium (Meq/100g)	Magnesium (Meq/100g)	Sulphur (mg kg^{-1})
S_1	0.76	1.18	0.19
S_2	3.46	2.86	0.25
S_3	4.90	0.82	0.20
S_4	3.10	0.80	0.32
S_5	1.42	0.12	0.34
S_6	0.94	0.24	0.23
S_7	0.96	0.04	0.31
S_8	2.40	0.50	0.45
S_9	5.62	2.18	0.25
S_{10}	3.42	0.24	0.26
S_{11}	1.96	0.78	0.25
S_{12}	4.00	0.90	0.14
S_{13}	6.60	1.24	0.24
S_{14}	7.54	0.44	0.16
S_{15}	9.08	0.66	0.20
S_{16}	1.14	0.58	0.18
S_{17}	2.56	1.56	0.27
S_{18}	0.86	1.10	0.15
S_{19}	3.00	3.70	0.41
S_{20}	3.46	1.24	0.43
S_{21}	0.40	0.56	0.18
S_{22}	1.88	1.68	0.25
S_{23}	1.16	0.08	0.29
S_{24}	0.42	1.72	0.26
MEAN	2.96	1.05	0.25
RANGE	0.40-9.08	0.04-3.70	0.14-0.45
SD±	2.34	0.89	0.08
CV (%)	79.1%	85.1%	32.6%

Appendix-6: Description of water sampling sites

Village name	Sample No.	Farmers name	Land type	Latitude	Longitude	Altitude
Site 1: Chipilli	S_1	Venkataraman	Cultivated tomato	13.5561**S**	78.48013**E**	709(±3m)
	S_2	Rambabu	Fallow land	13.55950**N**	78.48114**W**	717(±3m)
	S_3	Seena	Fallow land	13.56026**S**	78.48149**W**	721(±3m)
Site2: Ponnugutupalem	S_4	Ramireddy	Cultivated tomato	13.56047	78.48655**E**	710(±3m)
	S_5	Ravindranath	Fallow land	13.56194**N**	78.48207	720(±4m)
	S_6	Shiva reddy	Fallow land	13.56198**S**	78.48165	718(±3m)
Site:3 Nayabandapalli	S_7	Soma Shekar	Cultivated mango	13.6690**S**	78.4226**E**	718(±3m)
	S_8	Raja reddy	Grassed land	13.66880**S**	78.42162	721(±4m)
	S_9	Ravi chadra	Cultivated mango	13.67134**N**	78.42041	719(±3m)
Site:4 Kandlamadugu	S_{10}	S.A. Mastan	Fallow land	13.66824**N**	78.41873**W**	724(±3m)
	S_{11}	Pullaiah	Cultivated mango	13.66555**S**	78.41940**W**	725(±4m)
	S_{12}	Triveni	Cultivated mango	13.66466**N**	78.41672**E**	741(±3m)
Site:5 Basinikonda	S_{13}	Susheel	Fallow land	13.55242**N**	78.52216	680(±3m)

Village name	Sample No.	Farmers name	Land type	Latitude	Longitude	Altitude
	S_{14}	Naveen	Fallow land	13.55423**N**	78.51842**W**	676(±3m)
	S_{15}	Bharath	Fallow land	13.55285**N**	78.51864**E**	675(±3m)
Site:6 Ramacharlapallli	S_{16}	Siraj	Cultivated tomato	13.54999	78.53031**W**	676(±4m)
	S_{17}	Balu Swamy	Cultivated marigold	13.55068**S**	78.53369**E**	667(±4m)
	S_{18}	Ghouse	Fallow land	13.55101**S**	78.53814**W**	663(±3m)
Site:7 Ganganavaripalli	S_{19}	Venkatesh	Fallow land	13.54892	78.55058**W**	659(±3m)
	S_{20}	Vishnu	Fallow land	13.54892**S**	78.55048**E**	665(±3m)
	S_{21}	Chinnapa	Cultivated tomato	13.54969**S**	78.55230	671(±3m)
Site:8 Barlapalli	S_{22}	Shankar reddy	Cultivated gooseberry	13.54149	78.43281	744(±3m)
	S_{23}	Kallu Banda	Fallow land	13.54306	78.43305	748(±3m)
	S_{24}	Talla reddy	Fallow land	13.54706	78.43575	751(±3m)

Appendix-7: Status of available pH and EC in Water

Sample No	Water pH	Water EC (dSm^{-1})
W_1	6.21	0.783
W_2	6.59	1.136
W_3	7.07	1.228
W_4	6.68	1.15
W_5	7.38	1.071
W_6	6.98	1.008
W_7	6.94	0.84
W_8	7.07	0.937
W_9	6.95	0.947
W_{10}	7.95	0.615
W_{11}	6.70	0.504
W_{12}	6.48	1.082
W_{13}	6.80	3.464
W_{14}	6.95	3.366
W_{15}	6.98	3.321
W_{16}	6.83	2.747
W_{17}	6.86	3.936
W_{18}	6.80	4.328
W_{19}	8.22	1.304
W_{20}	7.25	1.585
W_{21}	6.65	1.124
W_{22}	8.34	0.967
W_{23}	6.21	0.587
W_{24}	7.08	0.876
MEAN	6.99	1.62
RANGE	6.21-8.34	0.504-4.328
SD±	0.53	1.17
CV (%)	7.61%	72.4%

Appendix-8: Status of available Cations in water.

Sample No	Calcium + Magnesium (Meq/lit)	Calcium (Meq/lit)	Potassium (Meq/lit)	Sodium (Meq/lit)	Magnesium (Meq/lit)
W_1	17.2	12.4	0	1.52	4.8
W_2	14.2	12.2	0.0512	2.08	2.0
W_3	28.2	15.8	0.0512	2.34	12.4
W_4	25.8	14.2	0.0512	2.04	11.6
W_5	20.6	20.2	0.0512	2.08	0.4
W_6	4.8	3.8	0.0512	1.95	1.0
W_7	22.8	15.2	0.0256	1.65	7.6
W_8	26.4	18.8	0.0256	1.73	7.6
W_9	6.6	6.2	0.0512	1.78	0.4
W_{10}	7.4	6.8	0	1.43	0.6
W_{11}	8.2	4.4	0	1.39	3.8
W_{12}	27.0	10.0	0.0256	1.82	17
W_{13}	58.4	17.8	0.0256	4.34	40.6
W_{14}	12.0	8.0	0.0512	4.34	4.0
W_{15}	58.2	15.8	0.0512	4.17	42.4
W_{16}	40.2	39.8	0.0256	3.30	0.4
W_{17}	13.4	11.8	0.0256	4.78	1.6
W_{18}	31.0	29.0	0.0512	4.78	2.0
W_{19}	16.4	8.6	0.0512	2.69	7.8
W_{20}	22.4	7.2	0.0256	3	15.2
W_{21}	5.8	5.2	0.0256	2.43	0.6
W_{22}	21.0	14.6	0.0512	2	6.4
W_{23}	24.6	10.0	0.0512	1.47	14.6
W_{24}	10.4	8.4	0.0256	1.65	2.0
MEAN	21.7	13.1	0.03	2.53	8.61
RANGE	4.8-58.4	3.8-39.8	0-0.512	1.39-4.78	0.4-42.4
SD±	14.4	8.15	0.01	1.13	11.3
CV (%)	66.1%	61.8%	51.6%	44.5%	131.8%

Appendix-9: Status of available anions in water

Sample No	Chloride (Meq/lit)	Carbonate (Meq/lit)	Bicarbonate (Meq/lit)
W_1	4.0	0	2.0
W_2	5.2	0	2.0
W_3	4.0	0	2.8
W_4	4.8	0	2.2
W_5	4.0	0	2.6
W_6	4.0	0	2.2
W_7	2.4	0	1.8
W_8	4.0	0	2.0
W_9	3.6	0	2.2
W_{10}	4.0	0	1.4
W_{11}	3.6	0	1.4
W_{12}	4.8	0	2.4
W_{13}	23.2	0	3.4
W_{14}	22.8	0	4.0
W_{15}	25.6	0	3.4
W_{16}	21.2	0	2.8
W_{17}	30	0	3.4
W_{18}	28	0	4.0
W_{19}	4.4	0	3.2
W_{20}	8.0	0	3.8
W_{21}	6.0	0	1.4
W_{22}	7.2	0	2.8
W_{23}	6.4	0	1.8
W_{24}	10.4	0	3.0
MEAN	10.06	0	2.58
RANGE	2.4-30	0	1.4-4
SD±	9.17	0	0.81
CV (%)	91.1%	0	31.3%

Appendix 10: Status of irrigation water quality parameters and index

Sample No.	SAR	PI	KR	SSP	RSC	IWQI
W_1	0.52	15.68	0.09	8.13	-15.20	129.3
W_2	0.78	21.50	0.15	12.81	-12.20	154.3
W_3	0.63	13.16	0.08	7.69	-25.4	491.7
W_4	0.57	12.67	0.08	7.34	-23.6	194.4
W_5	0.65	16.31	0.10	9.20	-18.0	373.7
W_6	1.26	50.91	0.41	28.96	-2.60	487.3
W_7	0.49	12.24	0.07	6.76	-21.0	181.1
W_8	0.48	11.21	0.07	6.18	-24.4	168.8
W_9	0.98	38.96	0.27	21.27	-4.40	115.07
W_{10}	0.75	29.63	0.19	16.24	-6.00	144.8
W_{11}	0.69	26.84	0.17	14.51	-6.80	168
W_{12}	0.50	11.71	0.07	6.33	-24.6	114.8
W_{13}	0.80	9.87	0.07	6.93	-55.0	93.4
W_{14}	1.77	38.83	0.36	20.60	-8.00	81.2
W_{15}	0.77	9.65	0.07	6.69	-54.8	176.4
W_{16}	0.76	11.44	0.08	7.60	-37.40	405.2
W_{17}	1.85	36.44	0.36	26.30	-10.00	442.9
W_{18}	1.21	18.96	0.15	13.37	-27.00	519.6
W_{19}	0.94	23.48	0.16	14.12	-13.20	173.28
W_{20}	0.90	19.49	0.13	11.81	-18.60	217.04
W_{21}	1.43	43.94	0.42	29.57	-4.40	130.7
W_{22}	0.62	15.97	0.10	8.70	-18.20	170.7
W_{23}	0.42	10.81	0.06	5.67	-22.80	137.5
W_{24}	0.72	28.08	0.16	13.71	-7.40	142.9
MEAN	0.85	21.9	0.16	13.1	-19.2	225.5
RANGE	0.42-1.85	9.65-50.9	0.06-0.42	5.67-29.5	-55.0 - -2.60	81.2-519.6
SD±	0.38	12.0	0.11	7.74	14.0	140.2
CV (%)	45.6%	54.9%	70.8%	58.7%	-73.3%	62.1%

Note: SAR=Sodium adsorption ratio, RSC=Residual sodium carbonate, KR=Kelly's ratio, SSP=Soluble sodium percentage, PI=Permeability index, IWQI=Irrigation water quality index.

Index

P

Q

P

R

S

T

V

W

9 789354 615184

Printed at Repro India Ltd.